Families of Bentham and Hooker Classification

DICOTYLEDONS

V. Darani M.Sc., M.Phil., SET

Chapter 1
Bentham and hooker's classification

Chapter 1: Bentham and Hooker Classification (1862-1883)

Division: Phanerogams or Seed Plants

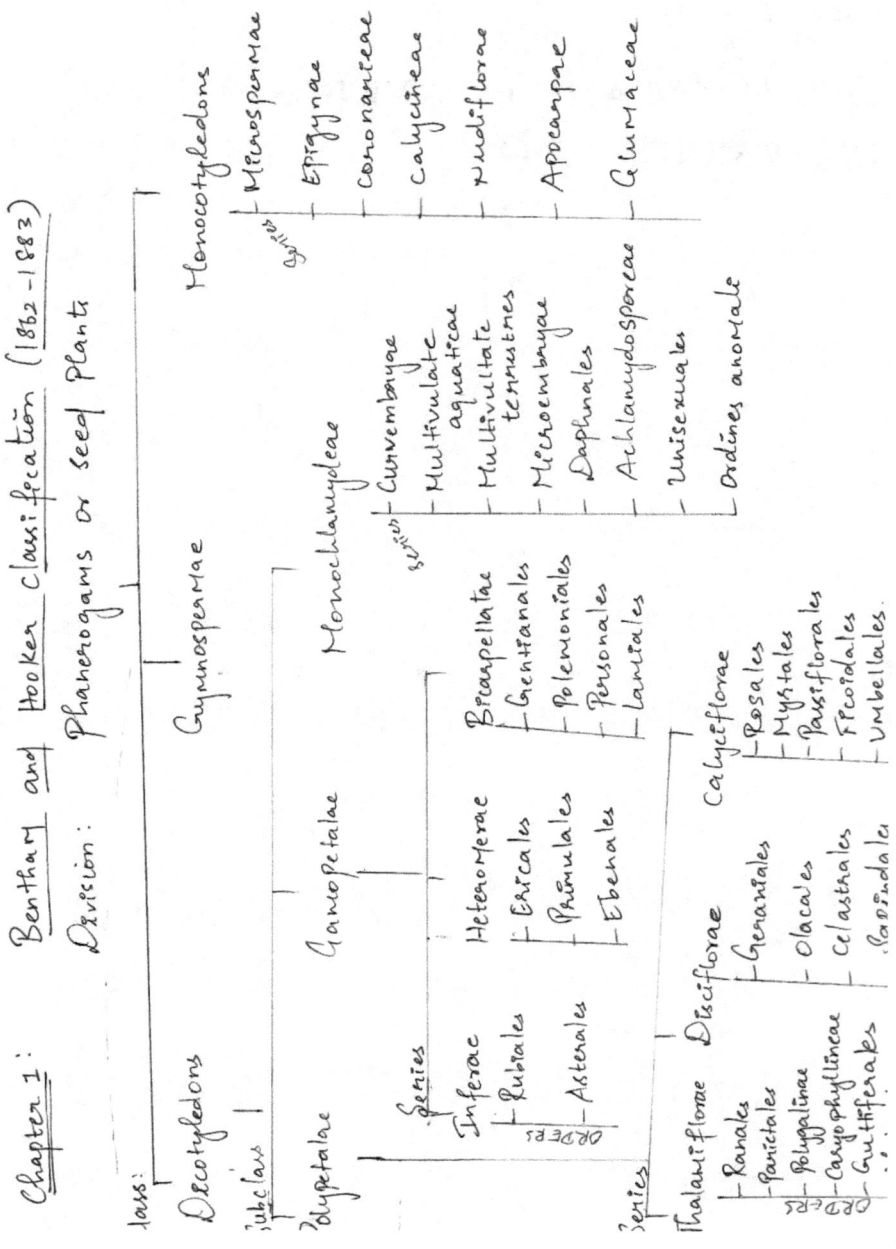

Families of Dicotyledons

Families listed in Bentham and Hooker's Classification

Subclass: Polypetalae
Series: Thalamiflorae
 Order: Ranales
  Families: 1) Ranunculaceae 2) Dilleniaceae 3) Calycanthaceae 4) Magnoliaceae 5) Annonaceae 6) Menispermaceae 7) Berberidaceae 8) Nymphaeaceae

 Order: Parietales
  Families: 1) Sarraceniaceae 2) Papaveraceae 3) Cruciferae 4) Capparidaceae 5) Resedaceae 6) Cistineae 7) Violaceae 8) Canellaceae 9) Bixineae

 Order: Polygalineae
  Families: 1) Pittosporeae 2) Tremandreae 3) Polygaleae 4) Vochysiaceae

 Order: Caryophyllineae
  Families: 1) Frankeniaceae 2) Caryophyllaceae 3) Portulacaceae 4) Tamariscineae

 Order: Guttiferales
  Families: 1) Elatineae 2) Hypericineae 3) Guttiferae 4) Ternstroemiaceae 5) Dipterocarpaceae 6) Chlaenaceae

Order: **Malvales**
Family: 1) Malvaceae 2) Sterculiaceae 3) Tiliaceae

Series: **Disciflorae**
Order: **Geraniales**
Families: 1) Lineae 2) Humiriaceae 3) Malphigiaceae 4) Zygophyllaceae 5) Geraniaceae 6) Rutaceae 7) Simarubeae 8) Ochnaceae 9) Burseraceae 10) Meliaceae 11) Chailletiaceae

Order: **Olacales**
Families: 1) Olacineae 2) Ilicineae 3) Cyrilleae

Order: **Celastrales**
Families: 1) Celastrineae 2) Stackhousieae 3) Rhamneae 4) Ampelideae

Order: **Sapindales**
Families: 1) Sapindaceae 2) Sabiaceae 3) Anacardiaceae 4) Coriarieae 5) Moringeae

Series: **Calyciflorae**
Order: **Rosales**
Families: 1) Connaraceae 2) Leguminosae 3) Rosaceae 4) Saxifrageae 5) Crassulaceae 6) Droseraceae 7) Hamamelideae 8) Bruniaceae 9) Halorageae

Order: **Myrtales**
Families: 1) Rhizophoraceae 2) Combretaceae

Families of Dicotyledons

3) Myrtaceae 4) Melastomaceae 5) Lythrarieae
6) Onagrarieae

Order: <u>Passiflorales</u>
Families: 1) Samydaceae 2) Loaseae 3) Turneraceae
4) Passifloreae 5) Cucurbitaceae 6) Begoniaceae
7) Daticeae

Order: <u>Ficoidales</u>
Families: 1) Cactaceae 2) Ficoideae

Order: <u>Umbellales</u>
Families: 1) Umbelliferae 2) Araliceae 3) Cornaceae

Subclass: <u>Gamopetalae</u>

Series: <u>Inferae</u>
Order: <u>Rubiales</u>
Families: 1) Caprifoliaceae 2) Rubiaceae

Order: <u>Asterales</u>
Families: 1) Valerianeae 2) Dipsaceae 3) Calycereae
4) Compositae.

Series: <u>Heteromerae</u>
Order: <u>Ericales</u>
Family: 1) Ericaceae 2) Vaccinieae 3) Monotropeae
4) Epacrideae 5) Diapensiaceae 6) Lennoaceae

Order: <u>Primulales</u>
Family: 1) Plumbagineae 2) Primulaceae
3) Myrsineae

Order: **Ebenales**
Family: 1) Sapotaceae 2) Ebenaceae 3) Styraceae

Series: **Bicarpellatae**
Order: **Gentianales**
Family: 1) Oleaceae 2) Salvadoraceae 3) Apocynaceae 4) Asclepiadaceae 5) Loganiaceae 6) Gentianaceae

Order: **Polemoniales**
Family: 1) Polemoniaceae 2) Hydrophyllaceae 3) Poragineae 4) Convolvulaceae 5) Solanaceae

Order: **Personales**
Family: 1) Scrophularineae 2) Orobranchaceae 3) Lentibularieae 4) Columelliaceae 5) Gesneraceae 6) Bignoniaceae 7) Pedalineae 8) Acanthaceae

Order: **Lamiales**
Family: 1) Myoporineae 2) Selagineae 3) Verbenaceae 4) Labiatae 5) Plantagineae

Subclass: **Monochlamydeae** (Incompletae)

Series: **Curvembryae**
Family: 1) Nyctagineae 2) Illecebraceae 3) Amarantaceae 4) Chenopodiaceae 5) Phytolaccaceae 6) Batideae 7) Polygonaceae

Series: **Multivulatae Aquaticae**
Family: 1) Podostemaceae

Series: **Multivulatae Terrestres**

Families of Dicotyledons

Family: 1) Nepenthaceae 2) Cytinaceae
3) Aristolochieae

Series: Micrœmbryae
Family: 1) Peperaceae 2) Chloranthaceae
3) Myristiceae 4) Moniaceae

Series: <u>Daphnales</u>
Family: 1) Laurineae 2) Proteaceae 3) Thymelaceae
4) Penaeaceae 5) Eleagnaceae

Series: Achlamydosporae
Family: 1) Loranthaceae 2) Santalaceae
3) Balanophoreae

Series: <u>Unisexuales</u>
Family: 1) Euphorbiaceae 2) Balanopseae
3) Urticaceae 4) Plantanaceae 5) Leitnerieae
6) Juglandeae 7) Myricaceae 8) Casuarineae
9) Cupuliferae

Series: Ordines <u>anomali</u>
Family: 1) Salicaceae 2) Lacistemaceae
3) Empetraceae 4) Ceratophylleae.

Monocotyledons

Series: Microspermae
Family: 1) Hydrocharideae 2) Burmanniaceae
3) Orchideae

Series: Epigynae
Family: 1) Scitamineae 2) Bromelliaceae 3) Haemodorac
4) Irideae 5) Amaryllideae 6) Taccaceae
7) Dioscoreaceae.

Series: Coronariae
Family: 1) Roxburghiaceae 2) Liliaceae
3) Pontederiaceae 4) Philydraceae 5) Xyrideae
6) Mayacaceae 7) Commelinaceae 8) Rapateaceae

Series: Calycineae
Family: 1) Flagellariaceae 2) Juncaceae 3) Palmae

Series: Nudiflorae
Family: 1) Pandaneae 2) Cyclanthaceae 3) Typhaceae
4) Aroideae 5) Lemnaceae

Series: Apocarpae
Family: 1) Triurideae 2) Alismaceae 3) Najadaceae

Series: Glumaceae
Family: 1) Eriocauleae 2) Centrolepideae
3) Restiaceae 4) Cyperaceae 5) Gramineae.

Sub class: polypetalae

Series: Thalamiflorae

Order: Ranales

Families:

Ranunculaceae

Dilleniaceae

Calycanthaceae

Magnoliaceae

Annonaceae

Menispermaceae

Berberidaceae

Nymphaeaceae

Chapter-2
Ranunculaceae

Classification (Bentham and Hooker)

- Phanerogams
- Dicotyledons
- Polypetalae
- Thalamiflorae
- Ranales
- Ranunculaceae

General characters:

→ Mostly herbaceous annuals or perennials, some woody climbers (*Clematis*) or shrubs (*Xanthorhiza*)

→ Flowers bisexual, showy or inconspicuous

→ May be solitary and are frequently found aggregated in various forms of inflorescence like cymes, panicles or spikes

→ Flowers — usually radially symmetrical.

→ Sepals, Petals, Stamens, carpels are generally free. The outer stamens may be modified to produce nectar (*Delphinium*).

→ In *Thalictrum*, sepals are colourful and appear like petals, whereas petals are inconspicuous or altogether absent

Families of Dicotyledons

→ Leaves → alternate or occasionally opposite or whorled. Most species possess both basal and cauline leaves, which are usually compound or lobed or simple.

→ Most perennials form rhizomes from which new roots arise each year.

→ Fruits — achenes or follicles or berry.

Genus included in Ranunculaceae

- Glaucidium
- Hydrastis
- Coptis
- Xanthorhiza
- Aquilegia
- Dichocarpum
- Enemion
- Isopyrum
- Leptopyrum
- Paraquilegia
- Paropyrum
- Semiaquilegia
- Thalictrum
- Urophysa
- Adonis
- Megaleranthis
- Ranunculus
- Trollius
- Aconitum
- Consolida
- Delphinium
- Nigella
- Helleborus
- Actaea
- Anemonopsis
- Beesia
- Cimicifuga
- Eranthis
- Soulieae
- Caltha
- Asteropyrum
- Callianthemum
- Anemoclema
- Anemone
- Clematis
- Trautvetteria
- Hepatica
- Naravelia
- Pulsatilla
- Barneoudia
- Calathodes
- Callianthemoides
- Ceratocephala
- Ficaria
- Halerpestes
- Hamadryas
- Knowltonia
- Krapfia
- Laccopetalum
- Metanemone
- Miyakea
- Oreithales
- Myosurus
- Oxygraphis
- Paroxygraphis

Chapter-3
Dilleniaceae

Classification (Bentham and Hooker)

    Phanerogams
    Dicotyledons
    Polypetalae
    Thalamiflorae
    Ranales
    Dilleniaceae.

General Characters:

→ Most of the members are woody plants, lianas or trees; but some herbaceous species are also noted (Pachynema).

→ Stem - self supporting or climbing, stem twiners in some cases.

→ Leaves - deciduous, alternate or rarely opposite, spiral, leathery or sometimes herbaceous or membranous, petiolate, sheathing or non sheathing, gland dotted; simple, entire or rarely lobed; one or pinnately or palmately veined; stipulate or exstipulate.

→ Flowers solitary or aggregated in inflorescence like cymose or racemose.

→ Flowers small to medium sized or rarely large.

→ Flowers regular to somewhat irregular
→ Perianth is with distinct calyx and corolla; Calyx – polysepalous, fleshy or non fleshy, persistent, spirally imbricate. Corolla – polypetalous, imbricate, white or yellow, deciduous. Petals – bilobed or entire
→ Androecium → 15–150 (or) 1–10 (rare), when numerous maturing centrifugally, free of the perianth, all ... equal, free of one another or united basally. Anthers – basifixed or adnate, introrse or latrorse
→ Gynoecium – Syncarpous or apocarpous when syncarpous (2-)5(-7) celled., Placentation Marginal or basal.
→ Fruit – non fleshy, aggregate or non aggregate, follicle or achene
→ Seeds endospermic and the endosperm is oily. Well differentiated embryo is present

Genus included in the family:

→ Acrotrema
→ Curatella
→ Davilla
→ Didesmandra
→ Dillenia
→ Doliocarpus
→ Hibbertia
→ Pachynema
→ Penzona
→ Schumacheria
→ Tetracera

Chapter-4
Calycanthaceae

Classification (Bentham and Hooker)

- Phanerogams
- Dicotyledons
- Polypetalae
- Thalamiflorae
- Ranales
- Calycanthaceae

General Characters:

→ Aromatic, deciduous shrubs; evergreen tree (Idiospermum)

→ Bear essential oils.

→ Leaves - opposite, leathery, petiolate, gland/dotted, aromatic, simple, entire, pinnately veined, exstipulate with entire margin.

→ Flowers - hermaphrodite and pollination is entomophilous.

→ Flowers - solitary, terminal, medium to large sized, acyclic. Receptacle is markedly hollow, free hypanthium is present, flowers - perigynous.

→ Perianth sequentially intergrading from sepals to petals, 15-30, free

→ Androecium → 15-55, members mature centripetally, free of the perianth, free of one another, spirally arranged on the top of the hypanthium. Staminodes are present and are 10-25 in number, internal to the fertile stamens, non petaloid, nectariferous; Stamens - 5-30, laminar or filantherous. Anthers - adnate, dehisce via longitudinal slits, extrose.

→ Gynoecium 5-45 carpelled, apocarpous, spirally arranged within hypanthium (eu-apocarpous; superior; Carpel stylate, the terminal style long and filiform with a decurrent stigma, 2 ovuled, placentation marginal.

→ Fruit - non fleshy, aggregate; an achene. Fruit enclosed in fleshy hypanthium.

→ Seeds non-endospermic with well differentiated embryo.

Genus belonging to family Calycanthaceae

→ Calycanthus
→ Chimonanthus
→ Idiospermum

Chapter-5
Magnoliaceae

Classification (Bentham and Hooker)

- Phanerogams
- Dicotyledons
- Polypetalae
- Thalamiflorae
- Ranales
- Magnoliaceae

General characteristics [Br $\oplus$ ⚥ P_{3+3+3} or $K_3, C_{3+3}, A_\infty, A_{\underline{\infty}}$]

→ Mostly trees or shrubs; few climbers

→ Oil Passages are found in Parenchyma of stem and leaves

→ Very Primitive anatomical characters are exhibited in woods.

→ Leaves: Simple, entire, alternate, stipulate. Stipules are quite large which unite to form a hood like structure protecting the leaves.

→ Cymose inflorescence

→ Flowers terminal or solitary axillary and large in size; hermaphrodite, actinomorphic, hypogynous, bracteate. Bracts often large and variously coloured.

→ Perianth is 3 whorled (sometimes somewhat different into calyx and corolla. Each whorl has 3 perianth

leaves
→ Androecium consists of numerous stamens arranged spirally.
→ Stamens – distinct and hypogynous. Anthers are bilobed and dehisce longitudinally.
→ Gynoecium consists of infinite number of carpels arranged spirally. Carpels are free (apocarpous). Gynoecium may be sessile or elongated. Number of ovules ranges from one to many.
→ Ovule – anatropous and in parietal placentation.
→ Fruit – follicle, winged nut (Samara) or berry.
→ Seeds endospermic with small embryo.
→ Pollinated by insects.

List of Genera included in Magnoliaceae

According to Rendle, there are 18 genera and 300 species in this family. Some of the genus are,

1. Manglietia
2. Magnolia
3. Lirianthe
4. Houpoea
5. Talauma
6. Oyama
7. Woonyoungia
8. Pachylarnax
9. Parakmeria
10. Alcimandra
11. Yulania
12. Michelia
13. Liriodendron

Chapter - 6
Annonaceae

→ Trees or Shrubs ; Some Climbers (Artabotrys, Uvaria)
→ Oil Passage Present in stems and flowers
→ cymose inflorescence
→ Mostly solitary. Sometimes Cauliflorous (borne on wood)
→ Flowers regular, large, hermaphrodite, bisexual, hypogynous (unisexual in Stelechocarpus)
→ Perianth consists of 3 whorls each with 3 or 2 segments. The outer whorl form the calyx and show valvate aestivation. Segments are free or somewhat connate at base. The two inner whorl constitute corolla (coloured) and show valvate or slightly imbricate aestivation.
→ Androecium with indefinite stamens arranged spirally on large convex receptacle above the perianth. Filaments - short and thick.
→ Anthers bicelled, introrse and the connective is elongated outside the anther.
→ Gynoecium consists of numerous free carpels which are arranged spirally on a large convex receptacle situated above the perianth axis.
→ Ovary superior, unilocular with numerous

anatropous ovules arranged in double row on the ventral suture of the carpel

→ Fruits – aggregate (etaerio) of berry. They are fleshy and form separate berries.
→ Seeds are large and endospermic with small embr.
→ Pollination by insects.

$\oplus \, \male \, P_{3+3+3}, A\infty, G\underline{\infty}$

Classification (Bentham and Hooker)

    Phanerogams
    Dicotyledons
    Polypetalae
    Thalamiflorae
    Ranales
    Annonaceae

List of Genus in Annonaceae

→ Afroguatteria
→ Alphonsea
→ Ambavia
→ Anaxagorea
→ Ancana
→ Annickia
→ Annona
→ Anonianthus
→ Anonidium
→ Artabotrys
→ Asimina
→ Asteranthe
→ Balonga
→ Bocagea
→ Bocageopsis
→ Boutiquea
→ Cananga
→ Cardiopetalum
→ Chieniodendron
→ Cleistopetalum
→ Cleistopholis
→ Craibella
→ Cremastosperma
→ Cyathocalyx
→ Cyathostemma
→ Cymbopetalum

Families of Dicotyledons

- → Dasoclema
- → Dasymaschalon
- → Dendrokingstonia
- → Dennettia
- → Desmopsis
- → Desmos
- → Diclinanona
- → Dielsiothamnus
- → Disepalum
- → Drepananthus
- → Duckeanthus
- → Duguetia
- → Ellipeia
- → Ellipeiopsis
- → Enicosanthum
- → Ephedranthus
- → Exellia
- → Fissistigma
- → Fitzalania
- → Friesodielsia
- → Froesiodendron
- → Fusaea
- → Goniothalamus
- → Greenwayodendron
- → Fusaea
- → Goniothalamus
- → Greenwayodendron

- → Guatteria
- → Haplostichanthus
- → Heteropetalum
- → Hexalobus
- → Hornschuchia
- → Isolona
- → Kinginda
- → Klarobelia
- → Letestudoxa
- → Malmea
- → Meiocarpidium
- → Meiogyne
- → Melodorum
- → Mezzettia
- → Mezzettiopsis
- → Mileusa
- → Mischogyne
- → Mitrella
- → Mitrephora
- → Monanthotaxis
- → Monocarpia
- → Monodora
- → Monoon
- → Mosannona
- → Neo-uvaria
- → Neostenanthera
- → Oncodostigma
- → Onychopetalum
- → Orophea
- → Oxandra

- → Oxymitra
- → Papualthia
- → Phaeanthus
- → Phoenicanthus
- → Piptostigma
- → Polyalthia
- → Polyceratocarpa
- → Popowia
- → Porcelia
- → Pseudannona
- → Pseudephedranth
- → Pseudomalmea
- → Pseudoxandra
- → Pseuduvaria
- → Raimondia
- → Richella
- → Rollinia
- → Ruizodendron
- → Sacropetalum
- → Sageraea
- → Sapranthus
- → Schefferomitra
- → Sphaerocoryne
- → Stelechocarpus
- → Stenanona
- → Tetrameranthu
- → Tridimeris
- → Trigynaea

- → Trivalvaria
- → Unona
- → Unonopsis
- → Uvaria
- → Uvariastrum
- → Uvariodendron
- → Uvariopsis
- → Xylopia
- → Xylopiastrum

Chapter-7
Menispermaceae

Classification (Bentham and Hooker)

- Phanerogams
- Dicotyledons
- Polypetalae
- Thalamiflorae
- Ranales
- Menispermaceae

General characters:

→ Mostly twining, lianas, rarely erect shrubs or small trees.
→ Branched tap root system
→ Stem is usually woody and twining and rarely erect
→ Leaf - Simple, Petiolate, exstipulate, mostly entire or occasionally Palmately lobed, mostly Palmately veined
→ Inflorescence is Racemose
→ Flowers - small, unisexual, greenish, generally actinomorphic, hypogynous, cyclic, trimerous or dimerous
→ Calyx - 6 Sepals arranged in 2 whorls of 3 each
→ Corolla - 6 Petals arranged in 2 whorls of 3 each and are usually smaller than that of sepals
→ Androecium - Staminate flowers with usually

Families of Dicotyledons

6 Stamens (sometimes 3-α), opposite to petals when of same number, free; variously connate or Monadelphous forming a central column (Cissampelos) anthers 4-celled, dehiscing longitudinally.

→ Gynoecium - Carpels 3 or more, in Pistillate flowers. apocarpous; Ovary superior; 1-loculed. ovules 2 aborting to 1, Parietal placentation, Style very short or absent, stigma terminal, capitate or discoid, entire or lobed.

→ Fruit - Drupe or achene

→ Endospermic or non endospermic seeds usually with curved embryo.

Genus belonging to the family Menispermaceae

- → Abuta
- → Albertisia
- → Anamirta
- → Anomospermum
- → Antizoma
- → Arcangelesia
- → Aspidocarya
- → Beirnaertia
- → Borismene
- → Burasaia
- → Calycocarpum
- → Carronia
- → Caryomene
- → Chasmanthera
- → Chlaenandra
- → Chondrodendron
- → Clonomene
- → Cissampelos
- → Cocculus
- → Coscinium
- → Curarea
- → Cyclea
- → Diplytheca
- → Dioscoreophyllum
- → Diploclisia
- → Disciphania
- → Echinostephia
- → Elephantomene
- → Eleutharrhena
- → Fibraurea
- → Haematocarpus
- → Hyperbaena
- → Hypserpa
- → Jateorhiza
- → Kolobopetalum
- → Legnephora

- Limacia
- Limaciopsis
- Macrococculus
- Menispermum
- Odontocarya
- Orthogynium
- Orthomene
- Pachygone
- Parabaena
- Perianthus
- Pericampylus
- Platytinospora
- Pleogyne
- Pycnarrhena
- Rhaptonema
- Rhigiocarya
- Sarcolophium
- Sarcopetalum
- Sciadotenia
- Sinomenium
- Sphenocentrum
- Spirospermum
- Stephania
- Strychnopsis
- Synandropus
- Synclisia
- Syntriandrium
- Syrrheonema
- Telitoxicum
- Tiliacora
- Tinomiscium
- Tinospora
- Triclisia
- Ungulipetalum

Chapter-8
Berberidaceae

Classification (Bentham and Hooker)

- Phanerogams
- Dicotyledons
- Polypetalae
- Thalamiflorae
- Ranales
- Berberidaceae

General characters

→ Perennial herbs and shrubs
→ Leaves - alternate, simple or pinnately compound (rarely palmately lobed), often spiny, exstipulate, some evergreen
→ Inflorescence - solitary, raceme, cyme or fasciculate (congested in clusters)
→ Flowers - bisexual, actinomorphic
→ Calyx - 3 to 15, separate (sepals)
→ Corolla - 3 to 9 petals or absent, separate
→ Androecium - stamens opposite petals, flattened in two series (biseriate), anthers open at slits or with flap-like valves.
→ Gynoecium consists of 2-3 carpels, united, hypogynous (superior)

→ Fruit – berry or capsule

Genus belonging to species Berberidaceae

- → Achlys
- → Alloberberis
- → Berberis
- → Bongardia
- → Caulophyllum
- → Diphylleia
- → Dysosma
- → Epimedium
- → Gymnospermium
- → Jeffersonia
- → Leontice
- → Mahonia
- → Moranothamnus
- → Nandina
- → Podophyllum
- → Ranzania
- → Sinopodophyllum
- → Vancouveria.

Chapter-9
Nymphaeaceae

Classification (Bentham and Hooker)

- Phanerogams
- Dicotyledons
- Polypetalae
- Thalamiflorae
- Ranales
- Nymphaeaceae

General characters:

→ Aquatic herbs
→ Stem is a rhizome which is short, thick and erect
→ Stem may survive for few years (*Victoria*) or for one year only (*Euryale*)
→ Leaves – petiolate, simple, floating and large in size. Peltate or ovate in shape. Heterophylly prevails in some species (*Cabomba*), where the floating leaves are ovate or peltate and the submerged leaves are finely dissected.
→ Mostly the leaf surface is shiny and smooth and in rare cases the lower surface is prickly (*Victoria* & *Euryale*)
→ Inflorescence is of cymose type
→ Flowers are solitary and found on long pedicels.

→ Flower - large, showy, variously coloured, pedicellate, hermaphrodite, actinomorphic, complete, hypogynous to epigynous, acyclic, cyclic or hemicyclic.

→ Perianth consists of indefinite free perianth leaves and usually differentiated into calyx and corolla.

→ Calyx consists of 3-5 sepals or indefinite sepals.

→ Sepals are green in colour but in <u>Nuphar</u>, they become petaloid and are larger than petals.

→ Corolla consists of 3 to indefinite petals which are free (polypetalous). Usually the petals are large, coloured and showy except in <u>Nuphar</u> where they are small and scale like.

→ Androecium consists of 3 to indefinite number of free stamens. When they are few in number i.e., 3 to 6, their arrangement is cyclic, but when they are indefinite, the arrangement is acyclic. The anthers are bicelled and introrse dehisce by longitudinal slits.

→ Gynoecium consists of 3 to indefinite number of carpels. They may be apocarpous or syncarpous. Ovary may be superior or inferior. Ovary is unilocular having many ovules found in parietal placentation.

→ Fruits are variable; may be follicles, berries, etaerio of indehiscent nutlets
→ Seeds are endospermic with well developed embryo possessing fleshy cotyledons. Seeds are often arillate.

Genus belonging to family Nymphaeaceae
→ Barclaya
→ Euryale
→ Nuphar
→ Nymphaeae
→ Victoria.

Sub class: polypetalae

Series: Thalamiflorae

Order: parietales

Families:

Sarraceniaceae

Papaveraceae

Cruciferae

Capparidaceae

Resedaceae

Cistineae

Violaceae

Canellaceae

Bixineae

Chapter-10
Sarraceniaceae

Classification (Bentham and Hooker)
- Phanerogams
- Dicotyledons
- Polypetalae
- Thalamiflorae
- Parietales
- Sarraceniaceae

General Characters:

→ Family of Pitcher plants.
→ Carnivorous plants that lure insects with nectar and use their elongated, tube shaped leaves filled with water and digestive enzymes to catch and consume them.
→ Perennial herbs
→ Leaves - tubular, vase like that hold water to drown insects. Brilliant colour and nectar help to attract insects in some species.
→ 4 to 5 sepals and 5 petals (Rarely 0) are present.
→ Stamens are numerous. Ovary superior and consists of 5 to 6 (rarely 3) united carpels (syncarpous) forming an equal number of chambers. It matures as a capsule.

• Genus belonging to Sarraceniaceae
→ Sarracenia → Darlingtonia → Heliamphora.

Chapter-11
Papavaraceae

Classification (Benthany and Hooker)

- Phanerogams
- Dicotyledons
- Polypetalae
- Thalamiflorae
- Parietales
- Papavaraceae

General characters

→ Mostly annual herbs, some undershrubs (Argemone), rarely shrubs (Dendromecon) and trees extremely rare (Boccionia)

→ Plants usually possess a milky or coloured latex

→ Branched tap root system is present

→ Stem is usually erect, herbaceous, cylindrical, branched, solid, green

→ Leaves- Simple, alternate, entire or more or less lobed or cut margins (Argemone), sometimes dissected

→ Inflorescence - cymose or Racemose. Flowers are arranged in compound raceme (Panicle) in Macbaya

→ Flower - Large, showy, attractive, bisexual, actinomorphic, complete, hypogynous.

Families of Dicotyledons

→ Perianth – biseriate or triseriate
→ Calyx consists of 2-3 deciduous sepals, usually polysepalous, aestivation twisted or imbricate
→ The number of petals is usually double the number of sepals. Usually the petals are arranged in two whorls and are variously coloured. Petals fall off quickly. Aestivation is imbricate.
→ Androecium is represented by indefinite stamens arranged in several alternating whorls. Polyandrous. Anthers bilobed, dehisce longitudinally, extrorse
→ Gynoecium consists of 2 to indefinite carpels, syncarpous. Ovary is superior and unilocular. Parietal placentation. Ovules are anatropous.
→ Fruit is capsule or nut. Capsule opens by valves or pores.
→ Seeds are small and endospermic with a crested smooth raphe
→ Pollination occurs by the agency of insects.

Genus belonging to family Papavaraceae

→ Hypecoum
→ Pteridophyllum
→ Adlumia
→ Capnoides
→ Corydalis
→ Dactylicapnos
→ Dicentra
→ Ehrendorferia
→ Ichtyoselmis
→ Lamprocapnos
→ Ceratocapnos
→ Cryptocapnos
→ Cysticapnos
→ Discocapnos
→ Fumaria

- → Fumariola
- → Rupicapnos
- → Sanguinaria
- → Platystemon
- → Meconopsis
- → Argemone
- → Platycapnos
- → Sarcocapnos
- → Eschscholzia
- → Chelidonium
- → Hunnemannia
- → Dendromecon
- → Pseudofumaria
- → Trigonocapnos
- → Stylophorum
- → Glaucium
- → Papaver

Chapter-12 Cruciferae (Brassicaceae)

Classification (Bentham and Hooker)

- Phanerogams
- Dicotyledons
- Polypetalae
- Thalamiflorae
- Parietales
- Cruciferae

→ The plants are mostly annual, biennial or perennial herbs. Rare cases shrubs.

→ Branched tap root. Tap root modifications are noted:— fusiform root (*Raphanus*), napiform root (*Brassica*).

→ Stem is herbaceous. In *B. oleraceae*, stem becomes corm like and very much thickened. In *Raphanus sativus*, stem is very much condensed.

→ Leaves — Simple, alternate, exstipulate and bear simple or branched hairs. They are usually radial or cauline, sub sessile or sessile, lyrate. When radial they are found in rosettes.

→ Inflorescence is generally of racemose type and very often may be raceme or corymb or corymb raceme. Bracts and bracteoles are absent.

→ Flowers — ebracteate, pedicellate, actinomorphic, hermaphrodite, complete, cruciform and

hypogynous.
→ Calyx consists of 4 sepals and is polysepalous. The sepals are arranged in 2 whorls of 2 each. The lateral inner sepals are sometimes pouched at the base serving as nectar containers. Imbric aestivation

→ Corolla is polypetalous and cruciform, consisting of four petals. Petals are found in one whorl. They are alternate to the sepals. Each petal is usually differentiated into a narrow claw and a broad expanded limb.

→ Androecium consists of 6 stamens arranged in 2 whorls. The outer two stamens are short whereas the inner four are long (tetradynamous). Nectaries are developed as small green gland at the base of the two short stamens. Anthers - bilobed, basifixed and introrse.

→ Gynoecium consists of 2 carpels (bicarpellary), syncarpous. Ovary - superior, unilocular but becomes bilocular due to the development of false septum or replum from the ingrowths of parietal placentas. Placentation is parietal. Many anatropous or camphylotropous ovules develop from the parietal placentae. Style is short with two lobed stigmas.

Families of Dicotyledons

→ Fruit is either siliqua or silicula; lomentum (*Raphanus*)

→ Pollination is entomophilous.

Genus belonging to family Cruciferae

- → Aethionema
- → Agallis
- → Alliaria
- → Alyssoides
- → Alyssopsis
- → Alyssum
- → Ammosperma
- → Anastatica
- → Anchonium
- → Andrzeiowskia
- → Anelsonia
- → Aphragmus
- → Aplanodes
- → Arabidella
- → Arabidopsis
- → Arabis
- → Arcyosperma
- → Armoracia
- → Aschersoniodoxa
- → Asperuginoides
- → Asta
- → Atelanthera
- → Athysanus
- → Aubrieta
- → Aurinia
- → Ballantinia
- → Barbarea
- → Beringia
- → Berteroa
- → Berteroella
- → Biscutella
- → Bivonaea
- → Blennodia
- → Boechera
- → Boleum
- → Boreava
- → Bornmuellera
- → Borodinia
- → Botscantzevia
- → Brachycarpaea
- → Brassica
- → Braya
- → Brayopsis
- → Brossardia
- → Bunias
- → Cakile
- → Calepina
- → Calymmatium
- → Camelina
- → Camelinopsis
- → Capsella
- → Cardamine
- → Cardaminopsis
- → Cardaria
- → Carinavalva
- → Carrichtera
- → Catadysia
- → Catenulina
- → Caulanthus
- → Caulostramina
- → Ceratocnemum
- → Ceriosperma
- → Chalcanthus
- → Chamira
- → Chartoloma
- → Cheesemania
- → Cheiranthus
- → Chlorocrambe
- → Chorispora
- → Christolea

Families of Dicotyledons

- → Chrysobraya
- → Chrysochamela
- → Cithareloma
- → Clastopus
- → Clausia
- → Clypeola
- → Cochlearia
- → Coelonema
- → Coincya
- → Coluteocarpus
- → Conringia
- → Cordylocarpus
- → Coronopus
- → Crambe
- → Crambella
- → Cremolobus
- → Crucihimalaya
- → Chrytospora
- → Cuphonotus
- → Cusickiella
- → Cycloptychis
- → Cymatocarpus
- → Cyphocardamum
- → Dactylocardamum
- → Degenia
- → Delphinophytum
- → Descurainia
- → Diceratella
- → Dichasianthus
- → Dictyophragmus
- → Didesmus
- → Didymophysa
- → Dielsiocharis
- → Dilophia
- → Dimorphocarpa
- → Diplotaxis
- → Diponia
- → Diptychocarpus
- → Dithyrea
- → Dolichorhynchus
- → Dontostemon
- → Douepea
- → Draba
- → Drabastrum
- → Drabopsis
- → Dryopetalon
- → Egria
- → Elburzia
- → Enarthrocarpus
- → Englerocharis
- → Eremobium
- → Eremoblastus
- → Eremodraba
- → Eremophyton
- → Ermania
- → Ermaniopsis
- → Erophila
- → Eruca
- → Erucaria
- → Erucastrum
- → Erysimum
- → Euclidium
- → Eudema
- → Eutrema
- → Euzomodendron
- → Farsetia
- → Fezia
- → Fibigia
- → Foleyola
- → Fortuynia
- → Galitzkya
- → Geococcus
- → Glaribraya
- → Glastaria
- → Glaucocarpum
- → Goldbachia
- → Gorodkovia
- → Graellsia
- → Grammosperma
- → Guillenia
- → Guiraoa
- → Gynophorea
- → Halimolobos
- → Harmsiodoxa
- → Hedinia

Families of Dicotyledons

- Heldreichia
- Helsophila
- Hemicrambe
- Hemilophila
- Hesperis
- Heterodraba
- Hirschfeldia
- Hollermayera
- Hormathophylla
- Hornungia
- Hornwoodia
- Hugueninia
- Hymenolobus
- Ianhedgea
- Iberis
- Idahoa
- Iodanthus
- Iodanthus
- Ionopsidium
- Irenepharsus
- Isatis
- Ischnocarpus
- Iskandera
- Iti
- Ivania
- Jundzellia
- Kernera
- Kremeriella
- Lachnocapsa
- Lachnoloma
- Leavenworthia
- Lepidium
- Lepidostemon
- Leptaleum
- Lignariella
- Lithodraba
- Lobularia
- Lonchophora
- Loxostemon
- Lunaria
- Lyocarpus
- Lyrocarpa
- Macropodium
- Malcolmia
- Mancoa
- Maresia
- Mathewsia
- Matthiola
- Megacarpaea
- Megadenia
- Menkea
- Menonvillea
- Microlepidium
- Microsysymbrium
- Microstigma
- Morettia
- Moricandia
- Moriera
- Morisia
- Murbeckiella
- Muricaria
- Myagrum
- Nasturtiopsis
- Nasturtium
- Neomartinella
- Neotchehatchewi
- Neotorularia
- Neuryrenia
- Neslia
- Nesocrambe
- Neuontobotrys
- Notoceras
- Notothlaspi
- Ochthodium
- Octoceras
- Olimarabidopsis
- Onuris
- Oreoloma
- Oreophyton
- Ornithocarpa
- Onychophragmus
- Otocarpus
- Oudneya
- Pachycladon

- Pachymitus
- Pachyphragma
- Pachypterygium
- Parlatoria
- Parodiodoxa
- Parolinia
- Parrya
- Parryodes
- Paysonia
- Pegaeophyton
- Peltaria
- Peltariopsis
- Pennellia
- Petiniotia
- Petrocallis
- Petroravenia
- Phlebolobium
- Phlegmatospermum
- Phoenicaulis
- Physaria
- Physocardamum
- Physoptychis
- Physorrhynchus
- Platycraspedum
- Polyctenium
- Polypsecadium
- Pringlea
- Prionotrichon
- Pritzelago
- Pseuderucaria
- Pseudoarabidopsis
- Pseudocamelina
- Pseudoclausia
- Pseudovesicaria
- Psychine
- Pterygiosperma
- Pterygostemon
- Pugionium
- Pycnoplinthopsis
- Pycnoplinthus
- Pyramidium
- Quezeliantha
- Quidproquo
- Raffenaldia
- Raphanorhyncha
- Raphanus
- Rapistrum
- Reboudia
- Redowskia
- Rhammatophyllum
- Rhizobotrya
- Ricotia
- Robeschia
- Rollinsia
- Romanschulzia
- Rorippella
- Rorippa
- Rytidocarpus
- Sameraria
- Sarcobrabe
- Savignya
- Scambopus
- Schimpera
- Schivereckia
- Schizopetalon
- Schlechteria
- Schoenocrambe
- Schouwia
- Scoliaxon
- Selenia
- Sibara
- Sibaropsis
- Silicularia
- Sinapidendron
- Sinapis
- Sisymbrella
- Sisymbriopsis
- Sisymbrium
- Smelowskia
- Sobolewskia
- Solms-laubachia
- Sophiopsis
- Sphaerocardamum
- Sophiopsis

41 | Families of Dicotyledons

- → Sphaerocardamum
- → Spirorhynchus
- → Spryginia
- → Staintoniella
- → Stanfordia
- → Stanleya
- → Stenopetalum
- → Sterigmostemum
- → Stevenia
- → Straussiella
- → Streptanthella
- → Streptanthus
- → Streptoloma
- → Stroganowia
- → Stubendorffia
- → Subularia
- → Succowia
- → Synstemon
- → Synthlipsis
- → Taphrospermum
- → Tauscheria
- → Teesdalia
- → Teesdaliopsis
- → Tetracme
- → Thellungiella
- → Thelypodiopsis
- → Thelypodium
- → Thlaspeocarpa

- → Thlaspi
- → Thysanocarpus
- → Trachystoma
- → Trichotolinum
- → Trochiscus
- → Tropidocarpum
- → Turritis
- → Vella
- → Warea
- → Weberbauera
- → Werdermannia
- → Winklera
- → Xerodraba
- → Yinshania
- → Zerdana
- → Zilla

Chapter-13
Capparidaceae

Classification (Bentham & Hooker)

- Phanerogams
- Dicotyledons
- Polypetalae
- Thalamiflorae
- Parietales
- Capparidaceae

General characters:

→ Herbs, shrubs or trees (*Capparis religiosa*); rarely climbers (*Maerua*)

→ Branched tap root system

→ Stem – Erect, branched, woody or herbaceous, solid, cylindrical

→ Leaves – Simple or palmately compound. Stipulate or exstipulate; in some cases stipules modified into spines (*C. spinosa*). Leaves are absent in some species (*C. aphylla*)

→ Inflorescence is racemose type. They may be arranged in raceme, corymbs, umbels (*C. seppani*) corymbose clusters or solitary (*Niebuhria*)

→ Flowers – Pedicellate and bracteate. Bracteoles are absent, hermaphrodite, actinomorphic, complete, regular or irregular, hypogynous.

Families of Dicotyledons

→ Calyx: Consists of four sepals arranged in two whorls of 2 each. The sepals are free (polysepalous). In *C. decidua*, sepals are unequal in size and the posterior sepals form a hood like structure. Aestivation - valvate or imbricate.

→ Corolla consists of four petals, polypetalous, usually regular, rarely irregular. Aestivation - valvate or imbricate.

→ Androecium consists of 4 to ∞ stamens and the number of stamens varies with species.

→ Characteristics of this family is the development of internode between petals and stamens (i.e) androphore or between stamens and pistil (gynophore). The androphore and gynophore are together called gynandrophore.

→ Filaments are filiform, anthers are basifixed and dithecous.

→ Gynoecium is situated on gynophore. It consists of 2 carpels (bicarpellary), syncarpous. In rare cases the carpel number is more than 2. The ovary is superior and unilocular; sometimes become multilocular by the ingrowth of parietal placentas. Placentation is parietal. Ovules are many and campylotropous. Style is very short

or absent.
→ Fruit is siliqua with replum, berry or drupe or nut.
→ Seeds are non endospermous, reniform and each seed contains a large variously folded embryo.
→ Pollination is entomophilous.

Genus in the family Capparidaceae

- Anisocapparis
- Apophyllum
- Bachmannia
- Belencita
- Boscia
- Buchholzia
- Cadaba
- Calanthea
- Capparicordis
- Capparidastrum
- Capparis
- Cladostemon
- Colicodendron
- Crateva
- Cynophalla
- Dhofaria
- Dipterygium
- Euadenia
- Hispaniolanthus
- Maerua
- Mesocapparis
- Monilicarpa
- Morisonia
- Neocalyptrocalyx
- Neothorelia
- Poilanedora
- Puccionia
- Quadrella
- Ritchiea
- Sarcotoxicum
- Steriphoma
- Thilachium
- Forchhammeria
- Haptocarpum
- Koeberlinia
- Oxystylis
- Pentadiplandra
- Podandrogyne
- Polanisia
- Setchellanthus
- Stixis
- Tirania
- Wislizenia

Excluded genera
- Borthwickia
- Cleome
- Cleomella
- Dactylaena

Chapter-14
Resedaceae

Classification (Bentham and Hooker)

- Phanerogams
- Dicotyledons
- Polypetalae
- Thalamiflorae
- Parietales
- Resedaceae

General characters:

→ Herbs (mostly) or few shrubs. Annual, biennial or perennial

→ Leaves - alternate, spiral, petiolate to sessile, non-sheathing, simple, epulvinate, lamina entire or dissected, pinnately veined; stipulate. Stipules - intrapetiolar, free of one another represented by glands

→ Plants usually hermaphrodite.

→ Flowers aggregate in inflorescence forms like racemes and spikes

→ Flowers - bracteate, small to medium sized, very irregular, zygomorphic, floral receptacle developing an androphore or gynophore or both androphore and gynophore. Free hypanthium absent.

- Perianth with distinct calyx and corolla or sometimes petals are absent (sepaline). Calyx 4-8, in 2 or 1 whorls, polysepalous or gamosepalous, unequal, corolla - (2-8) in 1 whorl, polypetalous, valvate or with open aestivation, white or yellow, petals clawed, fringed or deeply bifid.
- Androecium 3-50, mature centrifugally, free of the perianth, markedly unequal or rarely equal, free of one another or connate (olegomeris). Stamens 3-50, isomerous with perianth to polystemonous and anthers dehisce via longitudinal slits, introrse.
- Gynoecium 2-7 carpelled, Pistil when syncarpous is 1 celled. Gynoecium apocarpous to syncarpous. Ovary superior, placentation basal. Ovary - syncarpous and unilocular.
- Fruit - fleshy or non fleshy, aggregate or non aggregate, berry / follicle
- Seeds more or less endospermic with well differentiated embryo.

<u>Genus belonging to family Resedaceae</u>

- Borthwickia
- Caylusea
- Forchhammeria
- Homalodiscus
- Neothorelia
- Ochradenus
- Oligomeris
- Randonia
- Reseda
- Sesamoides
- Stixis
- Terania

Chapter-15: Cistineae

Classification: (Bentham and Hooker)
- Phanerogams
- Dicotyledons
- Polypetalae
- Thalamiflorae
- Parietales
- Cistineae

General characters:
→ Commonly known as rocknose
→ Bear Resin
→ Leaves - evergreen, opposite, simple, usually slight rough surfaced, 2-8 cm long
→ In some species leaves are coated with a highly aromatic resin called labdanum.
→ Showy 5 petaled flowers ranging from white to purple and dark pink., in a few species with a conspicuous dark red spot at the base of each petal

Genus belonging to family Cistineae
→ Cistus
→ Halimium
→ Helianthemum
→ Tuberaria

Chapter-16: Violaceae

Classification (Bentham and Hooker)

- Phanerogams
- Dicotyledons
- Polypetalae
- Thalamiflorae
- Parietales
- Violaceae

General Characters:

→ Annual or perennial herbs; under shrubs or shrubs are also found (*Alsodeia roxburghii* & *A. lanceolata*), *Alsodeia maingayi* is a small tree

→ Leaves – simple and stipulate, scattered or rarely opposite.

→ Inflorescence – cymose, rarely racemose. Sometimes two flowers arise in each axel.

→ Flowers – bracteate and bracteolate, hermaphrodite and usually zygomorphic, but rarely actinomorphic, complete and hypogynous. Pentamery prevails below gynoecium, there are 5 sepals, 5 petals, 5 stamens arranged in alternating whorls. Parts of these whorls are usually free, but sometimes connate at the base

→ Calyx consists of 5 sepals, sepals are free, green in colour and generally uniform in size

Families of Dicotyledons

- Sepals are persistant; imbricate aestivation.
- → Corolla consists of 5 petals; petals are variously coloured; Usually they are large and either equal or unequal in size; polypetalous; aestivation - imbricate or contorted.
- → Androecium consists of 5 stamens that are hypogynous. They are alternating with petals. The anthers are erect and form ring like structure around the style. The anthers are introrse, two celled and dehisce longitudinally. Anthers bear apical appendages. Stamens possess short filaments.
- → Gynoecium consists of 3 to 5 carpels. Ovary is unilocular and superior. Parietal placentation. Ovules are many, anatropous. Style - simple, short and terminal. The stigma is very often hooded.
- → Fruit is a capsule usually splitting into three boatshaped valves. Rarely fruit is a berry.
- → Seed is small, hard and shiny. They are endospermic and possess straight embryo. Seeds are winged.
- → Pollination is entomophilous.

Genus belonging to family Violaceae

- → Fusispermum
- → Leonia
- → Hymenanthera
- → Melicytus
- → Isodendrion
- → Amphirrhox
- → Paypayrola
- → Allexis
- → Decorsella
- → Gloeospermum
- → Rinorea
- → Rinoreocarpus

- Anchietea
- Corynostyles
- Hybanthopsis
- Hybanthus
- Mayanaea
- Noisettia
- Orthion
- Schweiggeria
- Viola.

Families of Dicotyledons

Chapter-17
Canellaceae

Classification (Bentham & Hooker)

    Phanerogams
    Dicotyledons
    Polypetalae
    Thalamiflorae
    Parietales
    Canellaceae

General characters:

→ Mostly trees rarely shrubs and are evergreen and glabrous
→ Bark is aromatic with prominent and unusual appearing lenticels.
→ Leaves — alternate, spiral or distichous, simple, entire, coriaceous, petiolate, pinnately veined, exstipulate with translucent (pellucid) glands.
→ Inflorescence are terminal or axillary, in a panicle (canella) or raceme; flowers are solitary in some species
→ Flowers — actinomorphic, hypogynous, trimerous. Receptacles are barely excavated and hypogynous disc is absent
→ Calyx consists of 3 (rarely 2) sepals which are thick, coriaceous and imbricate.

→ Corolla consists of (4)-5-12 petals arranged in 1-2(4) unlike whorls or spirally arranged, slender, imbricate in bud, usually free but in some cases connate at the base (_canella_) halfway to the apex (_cinnamosma_)

→ Androecium is monadelphous, adnate to the ovary. Number of stamens ranges from 6-12. Anthers- bithecous & extrorse, dehisce by longitudinal slits.

→ Gynoecium is syncarpous. Ovary has 6 carpels, unilocular and superior. The style is short and thick; stigma - apical and capitate with 2-6 lobes. Placentation is parietal. Ovules number range from 2 to many arranged in 1 or 2 rows

→ fruit - berry with persistent calyx and contain 2 or more seeds. Embryo is small and straight to slightly curved with 2 cotyledons.

Genus belonging to Canellaceae

→ Canella
→ Cinnamodendron
→ Cinnamosma
→ Pleodendron
→ Warburgia

Chapter-18
Bixineae (Bixaceae)

Classification (Bentham and Hooker)

Phanerogams
Dicotyledons
Polypetalae
Thalamiflorae
Parietales
Bixineae

General Characters:
→ Plants may be herbs, shrubs or trees
→ All plants of this family produce a red, orange or yellow latex
→ Leaves — simple, alternate
→ Flowers — actinomorphic, regular, symmetrical
→ Fruits are edible in some species and gums are obtained from few others

Genera belonging to Bixineae
→ Amoreuxia
→ Bixa
→ Cochlospermum
→ Diegodendron

Sub class: polypetalae

Series: Thalamiflorae

Order: polygalineae

Families:

Pittosporeae

Tremandreae

Polygaleae

Vochysiaceae

Chapter-19
Pittosporeae (Pittosporaceae)
(Bentham & Hooker)

Classification

Phanerogams
Dicotyledons
Polypetalae
Thalamiflorae
Polygalineae
Pittosporeae

General Characters:

→ Dioecious trees, shrubs or twining vines with coloured juice or non latciferous, bearing essential oils or resins

→ Leaves - persistent, alternate or sometimes whorled herbaceous or leathery, petiolate, non sheathing simple, entire, pinnately veined, exstipulate; Margin entire or serrate.

→ Plants hermaphrodite or rarely polygamomonoecio

→ Flowers solitary or aggregated in inflorescence - cymes or in corymbs. Flowers- bracteolate, small to medium sized, regular, 5merous, cyclic, free, hypanthium absent

→ Perianth with distinct calyx and corolla ~10 in 2 whorls, isomerous. Calyx 5, 1 whorled, polysepalous or gamosepalous (sometimes basally connate), regular not persistent, imbricate. Corolla 5 in 1 whorl

Polypetalous or gamopetalous, imbricate, regular, Petals sessile or clawed.

→ Androecium – 5, sometimes weakly, basally connate or free of one another, 1 whorled. Stamens 5, isomerous with perianth, oppositisepalous laminar or filantherous. Anthers dorsifixed or dorsifixed to basifixed, non-versatile.

→ Gynoecium 2(–5) carpelled, syncarpous, superior usually ovary is unilocular, rarely 2(–5) locular

→ Fruit – capsule or berry. Seeds with oily endosperm. Embryo tiny and well differentiated.

Genera belonging to family Pittosporaceae

→ Auranticarpa
→ Bentleya
→ Billardiera
→ Bursaria
→ Cheiranthera
→ Citriobatus
→ Hymenosporum
→ Marianthus
→ Pittosporum
→ Rhytidosporum

Chapter-20
Tremandreae

Classification (Bentham & Hooker)

- Phanerogams
- Dicotyledons
- Polypetalae
- Thalamiflorae
- Polygalineae
- Tremandreae

General Characters:

→ Small shrubs or herbs. Sometimes the principal photosynthesizing function transferred to stems.

→ Leaves well developed, or much reduced or absent. Leaves - Minute or small, alternate or opposite or whorled, flat or rolled, herbaceous or leathery or membranous, imbricate, petiolate to sessile, non sheathing, simple, Lamina entire, pinnately veined, exstipulate.

→ Plants hermaphrodite

→ Flowers - slender pedunculate, solitary or in inflorescence. When solitary - axillary. Flowers - bracteate, small, regular, (3-)4 merous or pentamerous cyclic, hypogynous disc may be present

→ Perianth with distinct calyx and corolla; 6 or 8 or 10 in 2 whorls, isomerous. Calyx (3-)4-5; 1 whorl, polysepalous or rarely gamosepalous; regular,

valvate. Corolla (3-)4-5 in 1 whorl, Polypetalous, regular, white or pink or purple.
→ Androecium 6 or 8 or 10. Androecial members branched or unbranched, free of the perianth, free from one another or coherent; 4 adelphous or 5 adelphous; 1 whorled. Stamens 6 or 8 or 10, diplostemonous, alternisepalous, opposite the corolla members, erect in bud, filantherous (wrt short filaments). Anthers - basifixed, non-versatile
→ Gynoecium - 2 carpelled. Pistil 2 celled; Gynoecium syncarpous, superior, ovary 2 locular, Gynoecium Median, stylate. Styles 1, apical. Placentation axile. Ovules 1-2(-5) per locule, pendulous, epitropous; with ventral raphe; usually arillate or rarely non-arillate; anatropous;
→ Fruit - non fleshy, dehiscent, loculicidal or septicidal capsule. Seeds endospermic, winged (through a twisted appendage) or wingless with straight embryo.

Genera belonging to Tremandreae

→ Platytheca
→ Tetratheca
→ Tremandra

Chapter - 21
Polygaleae

Classification (Bentham and Hooker)
Phanerogams
Dicotyledons
Polypetalae
Thalamiflorae
Polygalineae
Polygaleae

General characters:
→ Commonly known as Milkwort family
→ Plants often with milky juice in roots.
→ Low herbs
→ Leaves - simple, entire, dotted, exstipulate
→ Flowers - irregular
→ Sepals - 5, the inner large, petaloid
→ Petals - 3, the anterior one is larger

Genera belonging to Polygaleae

- → Acanthocladus
- → Ancylotropis
- → Asemeia
- → Badiera
- → Bredemeyera
- → Caamembeca
- → Comesperma
- → Epirixanthes
- → Gymnospora
- → Hebecarpa
- → Heterosamara
- → Hualania
- → Monnina
- → Muraltia
- → Phlebotaenia
- → Polygala
- → Polygaloides
- → Rhinotropis
- → Salomonia
- → Securidaca

Chapter - 22
Vochysiaceae

Classification (Bentham and Hooker)

- Phanerogams
- Dicotyledons
- Polypetalae
- Thalamiflorae
- Polygalineae
- Vochysiaceae

General Characters:

→ Trees, shrubs or lianas, rarely herbs, resinous
→ Plants green and photosynthesizing, self supporting or climbing.
→ Leaves - Persistent, opposite or whorled, alternate (rarely), leathery, simple, Lamina entire; stipulate or exstipulate, stipules when present, small; represented by glands in some cases
→ Plants hermaphrodite.
→ Flowers aggregate in 'inflorescence' in panicles.
→ Flowers - bibracteolate, very irregular, obliquely zygomorphic. Flowers cyclic; tetracyclic. Free hypanthium may or may not be present.
→ Perianth with distinct calyx and corolla. Calyx consists of 5 sepals in 1 whorl, gamosepalous

connate basally. Calyx lobes longer than the tube. Corolla when present consists of 1-3 or 5 petals in 1 whorls, polypetalous, imbricate or contorted; more or less unequal but not bilabia
→ Androecium 1-7, free of perianth, free of one another or may be coherent by connate filamen (diadelphous in such case); include staminodes 2-4. Stamens 1-4 — with a single fertile member; Anthers dehisce via longitudinal slits
→ Gynoecium monocarpellary or tricarpellary. [Carpels reduced in number relative to perianth to biorjerous with perianth. Stamens too reduced in number relative to the adjacent perianth]
→ Pistil 1 or 3 celled, Gynoecium - syncarpous, but sometimes pseudomonomerous; superior or inferior (when pseudomonomerous); Ovary unilocula Style 1, apical or lateral; stigma 1, small; Placentation is lateral to apical (when ovary inferior) and axile (superior ovary). 2 ovules in the single cavity
→ Fruit - non-fleshy, dehiscent or indehiscent; a capsule or samara.
→ Seeds usually non-endospermic and rarely

endospermic; conspicuously hairy or may not, often winged and rarely wingless
- Embryo straight and well differentiated.

Genera belonging to Vochysiaceae

→ Callisthene
→ Erisma
→ Erismadelphus
→ Korupodendron
→ Qualea
→ Ruizterania
→ Salvertia
→ Vochysia

Sub class: polypetalae

Series: Thalamiflorae

Order: caryophyllineae

Families:

Frankeniaceae

Caryophyllaceae

Portulacaceae

Tamariscineae

Chapter-23
Frankeniaceae

Classification (Bentham and Hooker)

- Phanerogams
- Dicotyledons
- Polypetalae
- Thalamiflorae
- Caryophyllineae
- Frankeniaceae

General Characters

→ Salt tolerant (halophytic) or drought tolerant (xerophytic) shrubs, sub shrubs or herbaceous plants.

→ Leaves – simple, opposite, generally small and somewhat heather-like and often with salt excreting glands in sunken pits.

→ Flowers – small, either solitary or borne in various kinds of cyme. Each flower has 4 to 7 sepals joined at the base into a tube; 4-7 overlapping petals, narrowed at base. Stamens are often arranged in 2 whorls of 3 each.

→ The ovary is made up of one to 4 carpels.

→ Fruit – capsule, enclosed in the persistent sepals.

→ Seeds have central embryo with considerable starchy endosperm on each side.

Families of Dicotyledons

Genera belonging to family Frankeniaceae
→ Frankenia

Chapter - 24
Caryophyllaceae

Classification (Bentham and Hooker)

- Phanerogams
- Dicotyledons
- Polypetalae
- Thalamiflorae
- Caryophyllineae
- Caryophyllaceae

General Characters:

→ Most of them are annual while some are perennial herbs.

→ Stem is erect, branched, green, herbaceous, solid and mostly swollen at the nodes.

→ Leaves - Simple, opposite, entire, exstipulate. Leaves sometimes have shortly connate perfoliate base (Dianthus). They are linear to lanceolate in shape.

→ Chief anatomical feature is the presence of Caryophyllaceous type of stomata which possess 2 subsidiary cells placed at right angles to the guard cells.

→ Inflorescence is cymose. Usually it is a dichasium which later becomes a dichasial cyme ending in a monochasial cyme (cincinnus inflorescence). Sometimes flowers are arranged in racemes. In some species, the flowers are solitary (Arenaria).

→ Flowers – Pedicellate, actinomorphic, usually hermaphrodite and pentamerous, but rarely unisexual or tetramerous; regular, complete and hypogynous.

→ Each typical flower bears 5 whorl – 5 sepals, 5 petals, 2 pentamerous whorl of stamens, 5 carpels, 5 styles and 5 double rows of ovules and sometimes five partition walls in the basal part of the ovary.

→ Calyx consists of 5 or rarely 4 sepals which may be free or united together into a tube. They are usually persistent with membranous margins. The aestivation is imbricate.

→ Corolla consists of 5 or rarely 4 petals which are free (polypetalous). Petals usually notched or sometimes bifid. Aestivation imbricate.

→ Androecium consists of 10 stamens. Sometimes, the stamen is reduced to 8, 5, 4, 3 or even 1. They are polyandrous, Obdeplostemonous (i.e.) stamens are arranged in 2 whorls of 5 each, the stamens of the outer whorl are seen opposite to the petals and of inner whorl alternate the petals. Stamens are hypogynous and rarely perigynous. Anthers are bilobed and dehisce longitudinally. Sometimes petaloid staminodes are present. In *Spergula*, the 5 or 10 stamens are found to be arranged on a perigynous disc.

→ Gynoecium consists of 2 to 5 carpels, syncarpous. Styles are free. The ovary is superior and unilocular. Ovules are many, campylotropous and arranged on a central column. Placentation is free central which is the most characteristic of the family. The number of carpels corresponds to the number of styles and stigmas.

→ Fruit generally an unilocular capsule. Some case the fruit may be achene or nut (Herniaria, Scleranthus)

→ Seeds are small, endospermic. Embryo is curved in the endosperm. Dispersed by censor mechanism.

→ Pollination is entamophilous.

Genera belonging to family Caryophyllaceae

- → Acanthophyllum
- → Achyronychia
- → Agrostemma
- → Allochrusa
- → Alsinedendron
- → Ankyropetalum
- → Arenaria
- → Bolanthus
- → Bolbosaponaria
- → Brachystemma
- → Bufonia
- → Cardeonema
- → Cerastium
- → Cerdia
- → Colobanthus
- → Cometes
- → Corrigeola
- → Cucubalus
- → Cyathophylla
- → Dianthus
- → Diaphanoptera
- → Dicheranthus
- → Drymaria
- → Drypis
- → Eremogone
- → Geocarpon
- → Gymnocarpos
- → Gypsophila
- → Habrosia
- → Haya
- → Heliosperma
- → Herniaria
- → Holosteum
- → Honckenya
- → Illecebrum

Families of Dicotyledons

- → Kabulia
- → Krauseola
- → Kuhitangia
- → Lepyrodiclis
- → Lochia
- → Loeflingia
- → Lychnis
- → Melandrium
- → Mesostemma
- → Microphyes
- → Minuartia
- → Moehringia
- → Moenchia
- → Myosoton
- → Ochotorophela
- → Ortegia
- → Paronychia
- → Pentastemonodiscus
- → Petrocoptis
- → Petrorhagia
- → Philippiella
- → Phrynella
- → Pinosia
- → Pirinia
- → Pleioneura
- → Plettkia
- → Pollichia
- → Polycarpaea
- → Polycarpon
- → Polytepalum
- → Pseudostellaria
- → Pteranthus
- → Pycnophyllopsis
- → Pycnophyllum
- → Reicheella
- → Sagina
- → Sanctambrosia
- → Saponaria
- → Schiedea
- → Scleranthopsis
- → Scleranthus
- → Sclerocephalus
- → Scopulophila
- → Selleola
- → Silene
- → Spergula
- → Spergularia
- → Sphaerocoma
- → Stellaria
- → Stipulicida
- → Thurya
- → Thylacospermum
- → Uebelinia
- → Vaccaria
- → Velezia
- → Wilhelmsia
- → Xerotia

Chapter-25
Portulacaceae

Classification:
Phanerogams
Dicotyledons
Polypetalae
Thalamiflorae
Caryophyllineae
Portulacaceae

General Characters:

→ Succulent herbs or infrequently soft shrubs.

→ Leaves - entire, alternate or opposite, usually fleshy and usually with scarious or hair like stipules

→ Flowers - bisexual, Perianth differentiated into calyx and corolla. Calyx consists of 2 distinct or basally connate sepals and 4-6 distinct or basally connate petals.

→ Stamens and Petals are isomerous, opposite and sometimes adnate or the stamens may number 2-4 times as many as the petals.

→ Gynoecium consists of single compound pistil of 2-3 carpels with a 2-5 branched style and a superior or half inferior ovary that has a single locule containing 1-many basal ovules

→ fruit - loculicidal capsule. (sometimes circumscissile

Families of Dicotyledons

Genera included in the family Portulacaceae

- Amphipetalum
- Anacampseros
- Avonia
- Calandrinia
- Calyptrotheca
- Ceraria
- Cistanthe
- Dendroportulaca
- Grahamia
- Lenzia
- Mona
- Neopaxia
- Portulaca
- Portulacaria
- Schreiteria
- Spraguea

Chapter - 26
Tamariscineae

Classification (Bentham & Hooker)

- Phanerogams
- Dicotyledons
- Polypetalae
- Thalamiflorae
- Caryophyllineae
- Tamariscineae

General Characters:

→ Undershrubs, bushes or small trees
→ Leaves - alternate, minute scale like
→ Flowers - Solitary or in spikes, or panicled racemes, regular usually bisexual
→ Sepals and petals each 5, rarely 4, imbricate, free or connate below
→ Stamens 5 to 10, numerous, inserted on an annular disc; anthers versatile
→ Ovary syncarpous, styles 2-5 free or connate or 3-5 sessile stigmas; ovules numerous.
→ Capsule 3-5 valved
→ Seeds erect winged or covered with down.

Genera included under Tamariscineae

→ Tamarix

Sub class: Polypetalae

Series: Thalamiflorae

Order: Guttiferales

Families:

Elatineae

Hypericineae

Guttiferae

Ternstroemiaceae

Dipterocarpeae

Chlaenaceae

Chapter - 27
Elatineae (Elatinaceae)

Classification (Bentham and Hooker)

Phanerogams
Dicotyledons
Polypetalae
Thalamiflorae
Guttiferales
Elatineae

General characters

→ Shrubs or herbs, annual or perennial
→ Hydrophytic and helophytic
→ Leaves submerged and emergent
→ Leaves - Opposite or whorled (rarely). Petiolate to sessile; simple, epulvinate. Lamina entire; linear to ovate or obovate; pinnately veined; attenuate to the base. Leaves stipulate. Stipules interpetiolar. Lamina margins entire, or crenate or serrate.
→ Plants hermaphrodite
→ Flowers - solitary or aggregated in inflorescence; when solitary, axillary, when aggregated in cymes.; small, regular; 2-5(-6) merous; cyclic; tetracyclic or pentacyclic. Free hypanthium absent

→ Perianth with distinct calyx and corolla; 4-10(-12) in 2 whorls, isomerous. Calyx consists of 2-5(-6) sepals in 1 whorls; Polysepalous or gamosepalous, regular, imbricate. Corolla consists of 2-5(-6) petals arranged in 1 whorls; Polypetalous, Imbricate, regular, Persistent.

→ Androecium 2-5(6) or 4-10(-12), androecial members free of the Perianth; all equal; free of one another; 2 whorled or 1 whorled. Stamens 2-5(-6) or 4-10(-12); isomerous with the Perianth or diplostemanous; when 1 whorled — oppositisepalous. Anthers dorsifixed; versatile; dehiscing via longitudinal slits.

→ Gynoecium 2-5(6) carpelled. Carpels isomerous with the Perianth. Pistil 2-5(-6) celled. Gynoecium syncarpous; synovarious; Superior; ovary 2-5(-6) locular. Styles 2-5(6), free, apical. Stigmas — capitate. Placentation axile. Ovules 15-50 per locule.

→ Fruit — non fleshy, dehiscent, a capsule. Seeds non-endospermic. Cotyledons 2, Embryo straight to curved.

Genera belonging to Elatineae
  → Bergia
  → Elatine

Chapter-28
Hypericineae

Classification (Benthary and Hooker)

- Phanerogams
- Dicotyledons
- Polypetalae
- Thalamiflorae
- Guttiferales
- Hypericineae

General characters:

→ Commonly called as St. John's wort family
→ Oblong ovate, pellucid-punctate leaves, thread like branches, and less slender brittle stems, with black dotted flower petals.
→ Annual, perennial herbs, subshrubs or shrubs.
→ Leaves – simple and entire, in opposite pairs; they are sometimes dotted with black or translucent glandular spots.
→ Inflorescence consists of a branched, flat-topped cluster, each flower being radially symmetrical with superior ovary.
→ Flowers – Sepals 4 or 5, which tend to persist; Petals – 4 or 5 usually yellow, sometimes dotted with black specks; Stamens numerous and

Present on long filaments; styles – 3 to 5, often fused at the base
→ The fruit is a dehiscent capsule
→ Black seeds

Genera included in the family Hypericineae
→ Hypericum
→ Triadenum
→ Thornea
→ Lianthus

Chapter-29
Guttiferae (Clusiaceae)

Classification (Bentham and Hooker)

- Phanerogams
- Dicotyledons
- Polypetalae
- Thalamiflorae
- Guttiferales
- Guttiferae

General Characters:

→ Trees, shrubs, herbs and lianas; lactiferous to with coloured juice; or non-lactiferous to without coloured juice.

→ Leaves – Mostly opposite but in some cases (Caraipa) alternate; when alternate, spiral. Herbaceous or leathery; Petiolate to sessile; Commonly gland dotted but in some species not gland dotted; Simple; epulvinate; Lamina – entire; Pinnately veined; Leaves – stipulate or exstipulate. Stipules when present – gland dotted.

→ Flowers – hermaphrodite or monoecious or dioecious or polygamomonoecious.

→ Flowers rarely solitary and commonly aggregated in inflorescence forms like cymes, umbels or panicles. Flowers – bracteolate (the two bracts often close up under the calyx and not clearly distinguished) or ebracteolate;

→ Perianth generally distinguished into calyx and corolla. Calyx – 2-6 sepals arranged in 1 whorl, basally gamosepalous or polysepalous. Calyx lobes somewhat longer than the tube. Imbricate aestivation; regular.

→ Corolla consists of 2-6-(14) petals arranged in 1 whorl, Polypetalous or sometimes connate basally; Imbricate or contorted; regular; yellow or white; Petals clawed or sessile.

→ Androecium usually many (20-100) or rarely few (3-4); When many, the androecial members mature centrifugally; free of the Perianth, often grouped into bundles, sometimes united into a tube; in some cases they may be free of one another. Anthers usually introrse and dehisce by longitudinal slits.

→ Gynoecium 1-13 carpelled or more.; Syncarpous, Superior, ovary 1(-3) locular. Styles 1 or 3; Stigma 1 or 3.

→ Fruit – fleshy or non fleshy, dehiscent or indehiscent, a capsule or a berry.

→ Seeds non endospermic, winged or wingless. Embryo rudimentary at the time of seed dispersal to well differentiated.

Genera included under Guttiferae

- Allanblackia
- Calophyllum
- Caraipa
- Chrysochlamys
- Clusia
- Clusiella
- Cratoxylon
- Dystovomita
- Eliaea
- Endodesmia
- Garcinia
- Haploclathra
- Harungana
- Havetia
- Havetiopsis
- Kayae
- Kielmeyera
- Lebrunia
- Lorostemon
- Mahurea
- Mammea
- Marila
- Mesua
- Montrouziera
- Moronobea
- Neotatea
- Oedematopus
- Pentadesma
- Pelosperma
- Platonia
- Poecloneuron
- Psorospermum
- Quapoya
- Santomasia
- Thornea
- Thysanostemon
- Tovomita
- Tovomitidium
- Triadenum
- Vismia

Chapter - 30
Ternstroemiaceae (Pentaphylacaceae)

Classification (Bentham and Hooker)

Phanerogams
Dicotyledons
Polypetalae
Thalamiflorae
Guttiferales
Ternstroemiaceae

General characters

→ Trees or shrubs
→ usually evergreen
→ Leaves - alternate, spiral, sometimes distichous, simple, petiolate, rarely sessile, often assymetrical, coriaceous, margin entire, serrate or rarely serrulate or crenulate, small deciduous bristle like glands often terminating each tooth. Venation pinnate, black gland dots present abaxially; exstipulate
→ Inflorescence - axillary, cymose, rarely racemose or fasciculate; rarely solitary (Ternstroemia).
→ Flower - actinomorphic, bisexual or unisexual and functionally dioecious with pistillode present in staminate flowers and appearing bisexual, pedicellate, bracteoles - 2.

→ Calyx with 5-6 sepals, imbricate, distinct or basally connate, setae occasionally bordering the sepal; Persistent in fruits.

→ Corolla consists of 5 petals, imbricate, distinct or slightly connate at the base.

→ Androecium consists of 10-50 stamens or more, staminodes present in female flowers, anthers basifixed, occasionally slightly connate at the base, adnate to the base of the corolla, often hirsute.

→ Gynoecium syncarpous, ovary superior and narrow apically or appears semi-inferior, carpels 1-3 locule. Style-1, stigma 2-5.

→ Fruit - a berry, indehiscent or a drupe, fleshy or dry. Persistent calyx and style.

→ Seed pendulous, can be numerous, aril present in Ternstroemia.

Genera included under Ternstroemiaceae

- Adinandra
- Anneslea
- Balthasaria
- Cleyera
- Eurya
- Euryodendron
- Freziera
- Killipiodendron
- Paranneslea
- Pentaphylax
- Symplococarpon
- Ternstroemia
- Ternstroemiopsis
- Visnea

Chapter - 31
Dipterocarpaceae

Classification (Bentham and Hooker)

  Phanerogams
  Dicotyledons
  Polypetalae
  Thalamiflorae
  Guttiferales
  Dipterocarpaceae

General Characters

→ Plants are trees and rarely shrubs. They are usually large trees with high erect stems.

→ Branched system of resin ducts usually lined with epithelium are found in almost all the members of the family. In many members - resin, balsam, camphor etc, are found in ducts.

→ Leaves - Simple, entire, coriaceous; stipules small and sometimes surround the internode. The petiole is locally thickened

→ Inflorescence - arranged in paniculate terminal or axillary spikes or racemes

→ Flower - axis may be broad, saucer-shaped or concave; bisexual, hypogynous, pentamerous, complete

→ Calyx consists of 5 sepals, free, unequal, imbricate or valvate; calyx segments are persistent winglike and enclose the fruit.

→ Corolla consists of 5 petals, imbricate, polypetalous

→ Androecium — Stamens 5, 10, 15 or indefinite. Stamens 15 (Hopea) in 3 whorls, of which those of the outer and inner whorls are antisepalous, the middle five are antipetalous. Filaments usually short, rarely long; anther lobes unequal, connective produced at the apex very often.

Gynoecium — Carpels usually 3, syncarpous, 3-loculed superior ovary, two or more ovules in each locule; axile placentation. Stigma almost always exceeds the stamens.

→ Fruit — may be leathery, woody indehiscent, nut-like, mostly surrounded by persistent winged calyx segments.

→ Seed is non-endospermic

Genera included under Dipterocarpaceae

- → Anisoptera
- → Cotylelobium
- → Dipterocarpus
- → Dryobalanops
- → Hopea
- → Marquesia
- → Monotes
- → Neobalanocarpus
- → Parashorea
- → Pseudomonotes
- → Shorea
- → Stemonoporous
- → Upuna
- → Vateria
- → Vateriopsis
- → Vatica

Chapter - 32
Chlaenaceae (Sarcolaenaceae)

Classification (Bentham and Hooker)

- Phanerogams
- Dicotyledons
- Polypetalae
- Thalamiflorae
- Guttiferales
- Chlaenaceae

General Characters:

→ Trees or shrubs
→ Leaves - Persistent, alternate, leathery or herbaceous, petiolate, non sheathing, simple, lamina entire, pinnately veined; Leaves - stipulate, stipules often large, intrapetiolar; free of one another; usually caducous; Lamina margins entire
→ Plants hermaphrodite.
→ Flowers aggregated in 'Inflorescence', in cymes or in umbels or in panicles.
→ Flowers regular. Perianth is distinguished into Calyx and corolla. Calyx consists of 3-5 sepals, Polysepalous, when 5, the outer members smaller than the inner, when four either one or three outer members smaller; imbricate.

→ Corolla consists of 5-6 petals arranged in 1 whorl; Polypetalous or gamopetalous (petals sometimes very slightly united at the base). Corolla lobes markedly longer than the tube. Corolla contorted.

→ Androecium 5-10 or many; Androecial members branched; usually maturing centrifugally; free of one another or sometimes fasciculate, the bundles of the filament weakly connate at the base; and free of the perianth; polystemonous or isomerous with the perianth to diplostemonous.; filantherous (the filaments slender); Anthers dorsifixed or basifixed, dehiscing via longitudinal slits; extrose or introrse; bilocular, tetrasporangiate

→ Gynoecium 1-5 carpelled; Carpels reduced in number relative to the perianth or isomerous with perianth or increased in number relative to the perianth. The pistil 1-5 celled. Gynoecium usually syncarpous and rarely monomerous; superior ovary. Style 1 attenuate from the ovary; Stigma 1; expanded usually lobed. Placentation basal to axile. Ovules 1-2-15 per locule.

→ Fruit: - non fleshy, indehiscent, achene (or) dehiscent capsule or nut. Capsules when dehisce loculicidal.

Families of Dicotyledons

→ Seeds with endosperm (fleshy or horny, starchy) or without endosperm. Embryo straight and well differentiated

Genera included under Chlaenaceae:

→ Eremolaena
→ Leptolaena
→ Medusella
→ Pentachlaena
→ Perrierodendron
→ Rhodolaena
→ Sarcolaena
→ Schizolaena
→ Xerochlamys
→ Xyloolaena

Sub class: Polypetalae

Series: Thalamiflorae

Order: Malvales

Families:

Malvaceae

Sterculiaceae

Tiliaceae

Chapter - 33
Malvaceae

Classification (Bentham and Hooker)
- Phanerogams
- Dicotyledons
- Polypetalae
- Thalamiflorae
- Malvales
- Malvaceae

General characters:
→ Plants are herbs, shrubs or Trees.
→ Branched Tap root system
→ Stem - erect, herbaceous or woody, branched, cylindrical, solid usually with stellate hairs. Mucilage sacs are also found
→ Inflorescence - cymose type, very rarely racemose. This may be solitary, axillary terminal or compound cyme
→ Flower - Pedicellate, bracteolate (bracteoles arranged in a whorl called epicalyx), actinomorphic, regular, hermaphrodite, complete, hypogynous and Pentamerous.
→ Calyx consists of 5 sepals, gamosepalous (i.e.) the sepals are united to each other, sepaloid, inferior, valvate aestivation

→ Corolla consists of 5 petals, polypetalous but the petals are inferior and the aestivation is twisted. Petals are variously coloured.

→ Androecium consists of indefinite stamens, stamens found arranged on a staminal tube, stamens - Monadelphous. Staminal tube remains united at the base to the petals, and therefore, the condition is known to be epipetalous. The filaments are short and the anthers are monothecous i.e. one celled and dorsifixed. The anthers dehisce transversely.

Stamens are derived from profuse branching of five antipetalous stamens. The outer whorl has been lost, but however, represented in *Hibiscus* by five staminodes.

→ Gynoecium consists of 5 to indefinite carpels (polycarpellary), syncarpous. Ovary is superior and multilocular. Each locule bears one to many ovules. Placentation is axile. Styles are united and stigmas are free. Style passes through the staminal tube. The number of stigmas is as much as the number of carpels.

→ Fruit - loculicidal capsule or schizocarp
→ Seed - buried in a hairy covering formed from the testa. Reniform or ovoid in shape. Endosperm scanty

Genera included under Malvaceae

- Abelmoschus
- Abutilon
- Abutilothamnus
- Acaulimalva
- Alcea
- Allosidastrum
- Allowissadula
- Althaea
- Alyogyne
- Anoda
- Anotea
- Asterotrichion
- Bakeridesia
- Bastardia
- Bastardiastrum
- Bastardiopsis
- Batesimalva
- Billieturnera
- Briquetia
- Callirhoe
- Calyculogygas
- Calyptraemalva
- Cenocentrum
- Cephalohibiscus
- Cienfuegosia
- Codonochlamys
- Corynabutilon
- Christaria
- Decaschistia
- Dendrosida
- Decellostyles
- Dirhamphis
- Eremalche
- Fioria
- Fryxellia
- Gaya
- Goethea
- Gossypioides
- Gossypium
- Gynatrix
- Hampea
- Helicteropsis
- Herissantia
- Hibiscadelphus
- Hibiscus
- Hochreutinera
- Hoheria
- Horsfordia
- Howittia
- Humbertianthus
- Humbertiella
- Iliamna
- Julostyles
- Jumelleanthus
- Kearnemalvastrum
- Kitaebela
- Kokia
- Kosteletzkya
- Krapovickasia
- Kydia
- Lagunaria
- Lavatera
- Lawrencia
- Lebronnecea
- Lecanophora
- Lopimia
- Macrostelia
- Malachra
- Malacothamnus
- Malope
- Malva
- Malvastrum
- Malviscus

- Malvella
- Megistostegium
- Meximalva
- Modeola
- Modeolastrum
- Monteiroa
- Napaea
- Nayariophyton
- Neobaclea
- Neobrittonia
- Nototriche
- Palaua
- Pavonia
- Peltaea
- Periptera
- Perrierophytum
- Phragmocarpidium
- Phymosia
- Plagianthus
- Radyera
- Rhynchosida
- Robinsonella
- Ropasinalva
- Senra
- Sida
- Sidalcea
- Sidastrum
- Sphaeralcea
- Symphyochlamys
- Tarasa
- Tetrasida
- Thespesia
- Urena
- Urocarpidium
- Wercklea
- Wissadula.

Chapter - 34
Sterculiaceae

Classification (Bentham and Hooker)
- Phanerogams
- Dicotyledons
- Polypetalae
- Thalamiflorae
- Malvales
- Sterculiaceae

General Characters:

→ Mostly trees (sometimes cauliflorous), shrubs or herbs, sometimes climbers

→ Leaves - simple, entire, rarely palmately lobed or palmately compound, alternate. Stipulate with generally caducous stipules

→ Inflorescence - complicated inflorescence

→ Flowers - hermaphrodite but sometimes unisexual by abortion (Sterculia); actinomorphic but rarely zygomorphic, pentamerous and hypogynous.

→ Calyx consists of 3-5 sepals which are somewhat connate at the base. Valvate aestivation.

→ Corolla - very often petals are reduced in size. Petals are absent in Sterculia & Cola

→ Androecium - Stamens are found to be arranged in two whorls. Stamens are found opposite the sepals (antisepalous) and represented by staminodes or altogether absent; the stamens of the inner whorl are found opposite the petals (antipetalous) are fertile. Filaments are more or less united to form a tube (monadelphous)

→ Gynoecium consists of 4-5 carpels, syncarpous, ovary superior, ovary is 4-5 loculed, carpel generally antipetalous and has two to numerous anatropous ovules; Axile placentation, Sometimes ovary is found to be carried up along the stamens above the petals by androgynophore, Styles are as many as the number of carpels which are either free or united together.

→ Fruits - dry, which very often separate into cocci°.

→ Seeds - endospermic, an embryo with two folded leaf like cotyledons.

→ Pollination is entomophilous.

Genera included under Sterculiaceae

- Acropogon
- Aethiocarpa
- Ambroma
- Astiria
- Ayenia
- Brachychiton
- Byttneria
- Cheirolaena
- Chiranthodendron
- Cola
- Commersonia
- Corchoropsis
- Cotylonychia
- Dicarpidium
- Dombeya
- Eriolaena
- Firmiana
- Franciscodendron
- Fremontodendron
- Gilesia
- Glossostemon
- Guazuma
- Guichenotia
- Hannafordia
- Harmsia
- Helicteres
- Helmiopsiella
- Helmiopsis
- Heritiera
- Hermannia
- Herrania
- Hildegardia
- Keraudrenia
- Kleinhovia
- Lasiopetalum
- Leptonychia
- Lysiosepalum
- Mansonia
- Maxwellia
- Megatritheca
- Melhania
- Melochia
- Neoregnellia
- Nesogordonia
- Octolobus
- Paradombeya
- Paramelhania
- Pentapetes
- Pterocymbium
- Pterospermum
- Pterygota
- Rayleya
- Reevesia
- Ruizia
- Rulingia
- Scaphium
- Scaphopetalum
- Seringia
- Sterculia
- Theobroma
- Thomasia
- Trichostephania
- Triplochiton
- Trochetia
- Trochetiopsis
- Uladendron
- Ungeria
- Waltheria

Chapter - 35
Tiliaceae

Classification (Bentham and Hooker)

    Phanerogams
    Dicotyledons
    Polypetalae
    Thalamiflorae
    Malvales
    Tiliaceae

General characters

→ Plants are usually trees or shrubs
→ Leaves - alternate, entire, dentate or lobed stipulate. Stipules serve as organs of protection of leaves and as the leaves unfold, they fall off.
→ Mucilage cells are found both in the pith and cortical regions of the plant
→ Inflorescence - cymose and often very complex
→ Flowers are regular, actinomorphic, hermaphrodite complete, Tetra or Pentamerous and hypogynous. Rarely unisexual flowers develop on monoecious plants
→ Calyx consists of 5 sepals, rarely 3 or 4 sepals, Polysepalous, sometimes gamosepalous (connate at base); aestivation - valvate.

→ Corolla consists of 5 petals, Polypetalous. Glands are found at the base of the petals. Aestivation is imbricate. Petals are usually coloured but sometimes green.

→ In Androecium, the number of stamens range from 10 to ∞. The stamens are quite free or connate at the base only. Stamens remain inserted at the base of the petals. In _Grewia_, they are raised above the corolla and are situated on an androgynophore. Anthers are 2 celled and basifixed. Dehisce by means of apical pores or longitudinal slits.

→ Gynoecium – number of carpels range from 2 to ∞. Ovary is superior and contains 2 to many loculus. One or more ovules are found in the inner angle. Axile placentation. Simple style bears a lobed or capitate stigma.

→ Fruit – dehiscent or indehiscent; In _Tilia_ fruit is a globose indehiscent nutlet having one or rarely 2 seeds. In other genera fruit is a loculicidal capsule.

→ Seeds are endospermic and the embryo is straight.

→ Pollination is entomophilous.

Genera included under Tiliaceae

- Ancestrocarpus
- Apeiba
- Asterophorum
- Berrya
- Brownlowia
- Burretiodendron
- Christiana
- Clappertonia
- Colona
- Corchorus
- Craigia
- Desplatsia
- Diplodiscus
- Duboscia
- Eleutherostyles
- Entelea
- Erinocarpus
- Glyphaea
- Goethalia
- Grewia
- Hainania
- Heliocarpus
- Hydrocaster
- Jarandersonia
- Luehea
- Lueheopsis
- Microcos
- Mollia
- Mortoniodendron
- Neotessmannia
- Pentace
- Pentaplaris
- Pseudocorchorus
- Schoutenia
- Sicrea
- Sparmannia
- Tahitia
- Tetralix
- Tilia
- Trichospermum
- Triumfetta
- Vasivaea
- Vinticena

Bombacaceae

Classification (Bentham and Hooker)
- Phanerogams
- Dicotyledons
- Polypetalae
- Thalamiflorae
- Malvales
- Bombacaceae

General Characters:
→ Plants are tall trees with thick trunks and spreading branches
→ Leaves – alternate, simply or palmately compound, deciduous, often with slime or mucilage cells and the vesture of stellate hairs or peltate scales; stipules caducous.
→ Inflorescence – cymose, solitary or fascicled in leaf axils or situated opposite a leaf.
→ Flowers – Actinomorphic; rarely zygomorphic, bisexual, large and showy, commonly bracteate, often appearing before the leaves; perianth subtended by an involucre.
→ Calyx consists of 5 sepals, distinct or basally connate, valvate aestivation

→ Corolla consists of 5 petals or occasionally absent, contorted in bud, imbricate or free
→ Androecium – stamens 5 or many, distinct or monadelphous, the anthers are often one celled (monothecous), dehiscing longitudinally, the pollen grains smooth; staminodes often present
→ Gynoecium consists of 2-5 carpels, connate in a superior 2-5 loculed ovary, with 2-many anatropous ovules in each locule; placentation axile; style 1, capitate or lobed, stigmas 1-5.
→ Fruit – a loculicidal capsule, sometimes an indehiscent pod or berry like.
→ Seed – smooth, sometimes embedded in a pith like tissue or in a woolly proliferation of the pericarp, occasionally arillate, the endosperm scant to absent
→ Pollination is entomophilous.

Genera included under Bombacaceae

- Adansonia
- Aguaria
- Bernoullia
- Bombacopsis
- Bombax
- Camptostemon
- Catostemma
- Cavanillesia
- Ceiba
- Chorisia
- Coelostegia
- Cullenia
- Durio
- Eriotheca
- Gyranthera
- Huberodendron
- Kostermansia
- Matisia
- Neesia
- Neobuchia
- Ochroma
- Pachera
- Patinoa
- Phragmotheca
- Pseudobombax
- Quararibea
- Rhodognaphalon
- Rhodognaphalopsis
- Scleronema
- Septotheca
- Spirotheca

Sub class: Polypetalae

Series: Discirlorae

Order: Geraniales

Families:

Lineae

Humiriaceae

Malpighiaceae

Zygophyllaceae

Geraniaceae

Rutaceae

Simarubeae

Ochnaceae

Burseraceae

Meliaceae

Chailletiaceae

Chapter-36
Linaceae

Classification (Bentham and Hooker)
- Phanerogams
- Dicotyledons
- Polypetalae
- Disciflorae
- Geraniales
- Linaceae

General Characters
→ Mostly annual herbs, sometimes shrubs, trees also occur.
→ Branched tap root system
→ Leaves - alternate or opposite, rarely whorled, simple, entire, stipulate or exstipulate (stipules sometimes intrapetiolar)
→ Inflorescence - cymose, dechasial cyme or cincinnus (sometimes appearing racemose)
→ Flower - Bisexual (hermaphrodite), actinomorphic, complete, hypogynous, mostly pentamerous
→ Calyx consists of 5 sepals, rarely four, the sepals may be free or connate at the base (rarely 4-parted with 3-fid lobes), imbricate aestivation.
→ Corolla consists of 5 petals, rarely four,

Petals contorted in bud, free, often clawed, aestivation - imbricate or contorted; arranged alternately to sepals, usually blue coloured, white or red, early deciduous.

→ Androecium consists of 5 stamens, alternately arranged with petals, filaments connate at the base and form a ring below the ovary, outside the ring nectar secreting glands are present, some cases 5 stamens are alternated with 5 or 10 tooth like staminodes, anthers dithecous (2 celled), introrse, dehisce by longitudinal slits.

→ Gynoecium consists of 5 carpels or 3-4 carpels, 5 locules or falsely 10-loculed by the intrusion of carpel midribs, 2 ovules per locule, ovules are pendulous and anatropous, placentation axile, ovary superior, pistil one; there are as many styles as many ovary locules, styles are free, filiform; stigma capitate

→ Fruit - drupe surrounded by persistent calyx or septicedal capsule

→ Seeds - Albuminous or exalbuminous; embryo straight, rarely incurved

→ Pollination is entomophilous, sometimes self pollination too occurs.

Genera included under Linaceae

- Anisadenia
- Cliococca
- Hesperolinon
- Linum
- Radiola
- Reinwardtia
- Sclerolinon
- Tirpitzia

Chapter - 37
Humiriaceae (Houmiriaceae)

Classification (Bentham and Hooker)

- Phanerogams
- Dicotyledons
- Polypetalae
- Disciflorae
- Geraniales
- Humiriaceae

General characters

→ Evergreen Trees
→ Flowers rather small and distinctive
→ Stamens are more or less fused in a tube and have prolongations at their apices
→ The fruit is a one or two seeded drupe with a ridged stone.
→ The wood of Humiria, can become beautifully scented after it is attacked by fungi, and then it can be used as incense.

Genera included under Humiriaceae

→ Duckesia
→ Endopleura
→ Humiria
→ Humiriastrum
→ Hylocarpa
→ Sacoglottis
→ Schistostemon
→ Vantanea.

Chapter-38
Malpighiaceae

Classification (Benthary and Hooker)

- Phanerogams
- Dicotyledons
- Polypetalae
- Disciflorae
- Geraneales
- Malphighiaceae

General Characters

→ lianas or trees

→ opposite leaves and lack teeth; hairs are single celled and T-shaped, and there are often conspicuous glands on the leaf or petiole.

→ Flowers - 4 or 5 sepals, often with large paired glands on their banks; the petals are strongly narrowed and stalked at the base and are crumpled in bud; 3 carpels each with a single ovule.

→ Pollination by bees which removes waxes from the sepal glands.

→ Sepals and filaments often persist at the base of the fruit, which is typically dry and with a variable number of wings.

→ Androecium consists of 10 distinct and basally connate stamens in two whorls but some or half of them are commonly reduced to staminode.

→ Style - 3, ovary superior, Axile placentation

Genera included under Malpighiaceae

- → Acmanthera
- → Acridocarpus
- → Adelphia
- → Aenigmatanthera
- → Alicia
- → Amorimia
- → Aspicarpa
- → Aspidopterys
- → Banisteriopsis
- → Barnebya
- → Blepharandra
- → Brachylophon
- → Bronwenia
- → Bunchosia
- → Burdachia
- → Byrsonima
- → Calcicola
- → Callaeum
- → Calyptostylis
- → Camarea
- → Carolus
- → Caucanthus
- → Christianella
- → Clonodia
- → Coleostachys
- → Cordobia
- → Cottsia
- → Diacidia
- → Dicella
- → Digoniopterys
- → Dinemagonum
- → Dinemandra
- → Diplopterys
- → Echinopterys
- → Ectopopterys
- → Excentradenia
- → Flabellaria
- → Flabellariopsis
- → Gallardoa
- → Galphimia
- → Gaudichaudia
- → Glandonia
- → Heladena
- → Henleophytum
- → Heteropterys
- → Hiptage
- → Hiraea
- → Janusia
- → Jubelina
- → Lasiocarpus
- → Lophanthera
- → Lophopterys
- → Madagasikaria
- → Malpighia
- → Malpighiodes
- → Mascagnia
- → Mcvaughia
- → Mezia
- → Microsteira
- → Mionandra
- → Niedenzuella
- → Peregrina
- → Peixotoa
- → Philgamia
- → Psychopterys
- → Pterandra
- → Ptilochaeta

Families of Dicotyledons

- Rhynchophora
- Ryssopterys
- Skoliopteris
- Spachea
- Sphedamnocarpus
- Stigmaphyllon
- Tetrapterys
- Thryallis
- Triaspis
- Tricomaria
- Triopterys
- Tristellateia
- Verrucularia
- Verrucularina

Chapter - 39
Zygophyllaceae

Classification (Bentham and Hooker)

Phanerogams
Dicotyledons
Polypetalae
Disciflorae
Geraniales
Zygophyllaceae

General Characters

→ herbs or shrubs and rarely trees
→ Leaves - opposite, usually pinnately divided
→ Largely adapted to desert conditions
→ Flowers - regular, bisexual, sepals 5, distinct and Petals 5, distinct, rarely 4. Stamens either 5, 10 or 15. Ovary superior. It consists of 5 carpels, syncarpous with the partition wall present, forming an equal number of chambers. It matures as a capsule with 2 or more seeds per cell or rarely drupe.

Genera included under Zygophyllaceae

→ Bulnesia
→ Guaiacum
→ Izozogia
→ Larrea
→ Pintoa
→ Porlieria
→ Morkillia
→ Sericodes
→ Viscainoa

- Seetzenia
- Balanites
- Kallstroemia
- Kelleronia
- Neoluederitzia
- Sisyndite
- Tribulopis
- Tribulus
- Augea
- Fagonia
- Melocarpum
- Roepera
- Tetraena
- Zygophyllum
- Metharme
- Plectrocarpa

Chapter 40: Geraniaceae

Classification (Bentham and Hooker)

- Phanerogams
- Dicotyledons
- Polypetalae
- Disciflorae
- Geraniales
- Geraniaceae

General Characters

→ Annual herbs or undershrubs and very rarely shrubs. (Dirachma)

→ Stems - often woody below in herbs. In many genera stems are rhizome or tuber like at the base. They are thick and fleshy.

→ Leaves - alternate, stipulate, rarely opposite. They are incised or palmitifid or incised upto the base or compound, rarely entire. In xerophytic species (Sarcocaulon) possesses small leaves metamorphosed into spines. Stipules are small.

→ Inflorescence - cymose or solitary or biclustered. Flowers arranged in umbels (Pelargonium)

→ Flowers - bracteate, bracteolate, bisexual,

actinomorphic, rarely zygomorphic, hypogynous, pentamerous, complete
→ Calyx consists of 5 sepals rarely 4 or 8, polysepalous, rarely gamosepalous, imbricate, rarely twisted, Posterior sepal sometimes rarely spurred (Pelargonium)
→ Corolla consists of 5 petals, rarely 4 or 8, polypetalous, imbricate, rarely twisted. Petals alternate to sepals
→ Androecium consists of 10 stamens in 2 whorls, obdiplostemonous, outer stamens shorter than inner. anthers dithecous, basifixed with small connectives. Filaments at the base are united, rarely free.
→ Gynoecium usually consists of 3-5 carpels, rarely 2 or 8, syncarpous, ovary superior, 3-5 locular, usually 2 ovules in each locule, mostly campylotropous, ovary terminates into a beaked structure at the apex; stigma ligulate; style long, axile placentation
→ Fruit Capsule, dehiscing into 3-5 one or more seeded mericarps. Each valve is twisted or rolled
→ Seeds - Non endospermic or with scanty endosperm, embryo straight or curved
→ Pollination - entomophilous.

Genera included under Geraniaceae

→ Geranium
→ Erodium
→ Monsonia
→ Pelargonium
→ Sarcocaulon

Chapter - 42
Simaroubaceae

Classification (Bentham and Hooker)
- Phanerogams
- Dicotyledons
- Polypetalae
- Disciflorae
- Geraniales
- Simaroubaceae

General Characters

→ Trees and shrubs producing characteristic triterpenoid lactones (Simaroubalides); without resin canals.

→ Leaves - alternate, spiral, petiolate, non-sheathing; non-gland dotted, simple or compound, pinnate or unifoliolate or ternate. Lamina pinnately veined; cross-venulate, exstipulate, margin entire.

→ Inflorescence - cyme, racemes, spikes, panicles or catkins; rarely solitary.

→ Flower - Unisexual flowers present, plants monoecious, or dioecious or polygamomonoecious. Gynoecium of male flowers pistillodial, or vestigial or absent. Flowers minute or small; regular, 3-5-8 merous, cyclic; when hermaphrodite - pentacyclic. Floral receptacle developing an androphore, or gynophore or neither androphore or gynophore. Usually hypogynous disc present and rarely absent; when present – extrastaminal.

→ Perianth with distinct calyx or corolla (or) rarely corolla absent. 6-10-16; 2 whorled or rarely 1 whorled; isomerous; calyx 3-5-8 in 1 whorl, gamosepalous or polysepalous; regular; usually imbricate rarely valvate. Corolla consists of 3-5-8 petals in 1 whorl; pentamerous; usually imbricate rarely contorted or valvate; regular.

→ Androecium consists of 3-16 stamens, free of the perianth, free of one another, usually 2 whorled rarely one whorled; stamens diplostemonous or rarely isomerous with perianth or triplostemonous to polystemonous. Anthers – dorsifixed or basifixed; versatile; dehiscing via longitudinal slits; introrse or rarely extrorse.

→ Gynoecium 1 carpelled or 2-8 carpelled; carpel number not reduced relative to the perianth, or isomerous with perianth; apocarpous to syncarpous superior; carpel apically stigmatic or with a lateral style or gynobasic style.

→ Fruit – fleshy or non fleshy; aggregate or non-aggregate; indehiscent – drupaceous, baccate or samaroid.

→ Seeds – non endospermic; embryo chlorophyllous, straight or curved.

Genera under Simaroubaceae

- Ailanthus
- Amaroria
- Brucea
- Castela
- Eurycoma
- Gymnostemon
- Hannoa
- Iridosma
- Laumoniera
- Leitneria
- Nothospondias
- Odyendea
- Perriera
- Picrasma
- Picrolemma
- Pierreodendron
- Quassia
- Samadera
- Simaba
- Simarouba
- Soulamea

Chapter - 43
Ochnaceae

Classification (Bentham and Hooker)

Phanerogams
Dicotyledons
Polypetalae
Disciflorae
Geraniales
Ochnaceae

General Characters

→ Mostly shrubs and small trees; few herbs
→ Leaves - simple, except pinnately compound in Krukoviella, in Juvenile plants often pinnately lobed; coriaceous and serrate, stipules present; Venation - Scalariform in appearance; Petioles short or absent, sometimes resemble a pulvinus.
→ Flowers mostly unisexual but in Polygamous species flowers are assessed as bisexual based only on morphology.
→ Sepals 3 to 5, often unequal sometimes accrescent.
→ Petals 4 or 5 or rarely 3, 6, 7 or 8, often contort, free or fused at the base only, sometimes reflexed over the sepals.

→ Fertile stamens 5 to 10 to numerous, rarely one. Filaments sometime persistent, sometimes narrowed near the anthers.

→ Anthers basifixed or slightly dorsifixed, usually dehiscing by one or two apical or subapical pores, sometimes latrorsely by longitudinal slits.

→ Staminodes often present, free or connate, sometimes petaloid, sometimes enveloping the fertile stamens.

→ Ovary superior, longitudinally ribbed. Carpels completely fused or nearly separate; 2-15 upto 25. Style apical or gynobasic.

→ Fruit sometimes winged, rarely a nut or drupe; usually a septicidal capsule.

→ Seeds albuminous or exalbuminous, winged or not. Seed coat often includes a layer of cristarque cells which are sclereids, each containing calcium oxalate crystals in the form of a druse.

Genera included under Ochnaceae

→ Medusagyne
→ Froesia
→ Quiina
→ Touroulia
→ Lacunaria
→ Testulea
→ Philacra
→ Luxemburgia
→ Lophira
→ Perissocarpa
→ Ochna
→ Elvasia
→ Campylospermum
→ Ouratea
→ Idertia
→ Brackenridgea
→ Rhabdophyllum

- → Blastemanthus
- → Godoya
- → Rhytidanthera
- → Krukoviella
- → Cespedesia
- → Fleurydora
- → Poecilandra
- → Wallacea
- → Neckia
- → Schuurmansia
- → Schuurmansiella
- → Euthemis
- → Tyleria
- → Adenarake
- → Indosinia
- → Sauvagesia.

Chapter-44
Burseraceae

Classification (Bentham and Hooker)

- Phanerogams
- Dicotyledons
- Polypetalae
- Disciflorae
- Geraniales
- Burseraceae

General characters

→ Trees or shrubs; characterized by resins that are present within the plant tissue from the vertical resin canals and ducts in the bark to the leaf veins.

→ Leaves - alternate, spiral and odd pinnately compound with opposite, frequently long petiolate, entire to serrate, pinnately veined, exstipulate.

→ Axillary inflorescences carry small, radial, unisexual flowers. Plants tend to be dioecious.

→ 4 or 5 faintly connate sepals, imbricate and 4 or 5 imbricate petals are present.

→ Stamens contain nectar discs, having distinct glabrous filaments that occur in one or two whorls and in number equaling or twice the number of petals, tricolporate pollen is contained within two locules of the anthers that open longitudinally along slits.

→ Gynoecium consists of 2-5 connate carpels, one style and one stigma that is head like to lobed. Each locule of the superior ovary has 2 ovules; Axile placentation, ovule- anatropous to campylotropo

→ Fruit - drupe
→ Seed - exalbuminous.

Genera included under Burseraceae

→ Bursera
→ Commiphora
→ Aucoumea
→ Beiselia
→ Boswellia
→ Triomma
→ Garuga
→ Ambilobea
→ Canarium
→ Dacryodes
→ Haplolobus
→ Pseudodacryodes
→ Rosselia
→ Santiria
→ Scutinanthe
→ Trattinnickia
→ Crepidospermum
→ Protium
→ Tetragastris

Chapter - 41
Rutaceae

Classification (Bentham and Hooker)
- Phanerogams
- Dicotyledons
- Polypetalae
- Disciflorae
- Geraniales
- Rutaceae

General Characters

→ Usually trees or shrubs. Some shrubs are climbing and xerophytic. *Ruta graveolens* - strong smelling herb.

→ Branched tap root system

→ Stem - erect, sometimes climbing, branched, woody or rarely herbaceous, cylindrical, solid, green when young, grey when becomes old

→ Leaves - simple or compound. They are usually alternate rarely opposite; exstipulate; dotted with glands containing volatile oils which give typical smell to the leaves; winged petioles

→ Inflorescence - Mostly cymose, rarely racemose. Some case solitary, axillary.

→ Flower - Pedicellate, complete, hermaphrodite, regular or rarely slightly zygomorphic; white or yellow, hypogynous. Disc is present beneath the ovary. Usually pentamerous, rarely tetramerous or trimerous. Sometimes unisexual (*Evodia*)

→ Calyx consists of 5 sepals, rarely 4 sepals. Usually polysepalous but in genera with zygomorphic flowers, the calyx becomes either tubular or united. Aestivation – imbricate or quincuncial.

→ Corolla consists of 5 petals, rarely 4; Corolla is polypetalous. In *Correa speciosa*, petals become united attaining campanulate shape. Petals - white, yellow or red. Petals inferior. Aestivation - imbricate

→ Androecium usually 10 or 8 stamens in obdiplostemonous condition. Sometimes the number of stamens decreases or increases to indefinite. In indefinite condition, they are arranged in irregular bundles (ie) polyadelphous. Filaments free. Anthers – bilobed, basifixed and introrse

→ Gynoecium consists of 3, 4 or 5 carpels. Carpels are fused together. Placentation axile, ovary is tri to multilocular. Each locule contains one or more anatropous ovules. *Feronia* – parietal placentation

→ Fruit – drupe, berry or capsule

→ Seeds are endospermic or non-endospermic. Each seed has a large straight very often curved embryo.

→ Pollination entomophilous.

Genera included under Rutaceae

- Acmadenia
- Acradenia
- Acronychia
- Adenandra
- Adiscanthus
- Aegle
- Aeglopsis
- Afraegle
- Agathosma
- Almeidea
- Amyris
- Angostura
- Apocaulon
- Aralropsis
- Asterolasia
- Atalantia
- Balfourodendron
- Balsamocitrus
- Barosma
- Bergera
- Boenninghausenia
- Boninia
- Bosistoa
- Bouchardatia
- Bouzetia
- Brombya
- Burkillanthus
- Calodendrum
- Casimiroa
- Choisya
- Chorilaena
- Citropsis
- Citrus
- Clausena
- Clymenia
- Cneoridium
- Coleonema
- Comptorella
- Correa
- Crowea
- Cusparia
- Decagonocarpus
- Decatropis
- Decazyx
- Dictamnus
- Dictyoloma
- Diosma
- Diphasia
- Diphasiopsis
- Diplolaena
- Drummondita
- Dutaillyea
- Empleurum
- Eremocitrus
- Eriostemon
- Erythrochiton
- Esenbeckia
- Euchaetis
- Euodia
- Euxylophora
- Evodiella
- Fagara
- Fagaropsis
- Feronia
- Feroniella
- Fortunella
- Galepea
- Geijera
- Geleznowia
- Glycosmis
- Halfordia
- Haplophyllum
- Helietta
- Hortia
- Ivodea
- Kodalyodendron
- Lelonema

- Leptothyrsa
- Limnocitrus
- Limonia
- Lubaria
- Lunasia
- Luvunga
- Maclurodendron
- Macrostylis
- Medicosma
- Megastigma
- Melicope
- Merope
- Meurellia
- Metrodorea
- Microcitrus
- Microcybe
- Micromelum
- Monanthocitrus
- Monnieria
- Mutxeantha
- Murraya
- Myrtopsis
- Naringi
- Naudinia
- Nematolepis
- Neobyrnesia
- Zeenia
- Neoraputia
- Nycticalanthus
- Oricia
- Oriciopsis
- Orixa
- Oxanthera
- Pamburus
- Paramignya
- Peltostigma
- Pentaceras
- Phebalium
- Phellodendron
- Philotheca
- Phyllosma
- Pilocarpus
- Pitavia
- Platydesma
- Pleiospermium
- Plethadenia
- Polyaster
- Poncirus
- Pseudeosma
- Psilopeganum
- Ptelea
- Raputia
- Raputiarana
- Zanthoxylum
- Rauia
- Raulinoa
- Ravenia
- Raveneopsis
- Rhadinothamnus
- Ruta
- Rutaneblina
- Sarcomelicope
- Sargentia
- Severinia
- Sheilanthera
- Sigmatanthus
- Skimmia
- Spathelia
- Spiranthera
- Stauranthus
- Swinglea
- Teclea
- Tetractomia
- Tetradium
- Thamnosma
- Ticorea
- Toddalia
- Toddaliopsis
- Tractocopevodia
- Triphasia
- Vepris
- Wenzelia
- Zieridium

Chapter-45
Meliaceae
(Bentham and Hooker)

Classification
- Phanerogams
- Dicotyledons
- Polypetalae
- Disciflorae
- Geraniales
- Meliaceae

General characters:

→ Mostly trees or shrubs. Wood of this family emit characteristic smell.

→ Branched tap root system

→ Stem – Erect, branched, solid, woody with characteristic smell.

→ Leaves – alternate, petiolate, pinnately compound or decompound, rarely simple and exstipulate

→ Inflorescence – cymose type. Many cases cymose panicles are present.

→ Flower – actinomorphic, hermaphrodite, rarely unisexual, regular, complete, rarely incomplete and hypogynous.

→ Calyx consists of 4 or 5 sepals, poly or gamosepalous when gamosepalous, the sepals are connate at the base. Sepals are small in size. Mostly imbricate aestivation rarely valvate.

→ Corolla consists of 4 or 5 petals, rarely 3-8 petals are also found, Polypetalous, rarely they are connate or adnate to the staminal tube. Imbricate, valvate or contorted aestivation.

→ Androecium consists of 8-10 stamens. Rarely stamen numbers range from 5 to indefinite. Stamens united and form a long or short staminal tube (monadelphous). Sometimes, filaments are united in such case the anthers are sessile on the staminal tube. Generally a disc is found in between ovary and stamens. Anthers are bilobed and dehisce by longitudinal slits.

→ Gynoecium consists of 2-5 carpels, syncarpous. Ovary - superior; 2-5 loculed; 1, 2 or more ovules in each locule; placentation is axile. Ovules are pendulous and anatropous. Style is simple with capitate or lobed stigma.

→ Fruit - Berry, capsule or rarely drupe
→ Seeds - Albuminous or exalbuminous and winged.
→ Pollination - Anemophily or entomophily

Genera included under Meliaceae

- Aglaia
- Amoora
- Anthocarpa
- Aphanamixis
- Astrotrichilia
- Azadirachta
- Cabralea
- Calodecarya
- Capuronianthus
- Carapa
- Cedrela
- Chisocheton
- Chukrasia
- Cipadessa
- Dysoxylum
- Ekebergia
- Entandrophragma
- Guarea
- Heckeldora
- Humbertioturraea
- Khaya
- Lansium
- Lepidotrichilia
- Lovoa
- Malleastrum
- Melia
- Munronia
- Naregamia
- Neobeguea
- Owenia
- Pseudobersama
- Pseudocarapa
- Pseudocedrela
- Pterorhachis
- Reinwardtiodendron
- Ruagea
- Sandoricum
- Schmardaea
- Soymida
- Sphaerosacme
- Swietenia
- Synoum
- Toona
- Trichilia
- Turraea
- Turraeanthus
- Vavaea
- Walsura
- Xylocarpus.

Chapter - 46
Chailletiaceae (Dichapetalaceae)

Classification (Bentham and Hooker)

- Phanerogams
- Dicotyledons
- Polypetalae
- Disciflorae
- Geraniales
- Chailletiaceae

General Characters

→ Small trees, shrubs, leanas
→ Leaves - alternate, spiral, herbaceous or leathery, petiolate, non sheathing, simple. Lamina entire, pinnately veined; cross-venulate, stipulate. Stipule intrapetiolar, caducous. Lamina margins entire.
→ Inflorescence - cymes and in fascicles.
→ Flowers - unisexual flowers may be present or absent; plants hermaphrodite or monoecious; small, regular or somewhat irregular, cyclic. Hypogynous disc present.
→ Perianth with distinct calyx and corolla. Calyx with 4-5 sepals in 1 whorl, polysepalous or gamosepalous, imbricate. Corolla consists of 4-5 petals usually 2-lobed or bifid, 1 whorled, polypetalous (usually) or rarely gamopetalous (with a basal tube).

corolla imbricate, regular or unequal but not bilabio Petals broadly clawed or sessile, deeply bifid or bilobed or entire (rarely)

→ Androecium consists of 4-5 stamens, free of the Perianth or sometimes epipetalous, free of one another or coherent; 1 whorled; stamens isomerous with the Perianth; oppositisepalous; filanthrous or with sessile anthers. Dehisce via longitudinal slits.

→ Gynoecium - 2 carpelled or 3-4 carpelled. Pistil 2 celled or 3-4 celled. Gynoecium syncarpous; superior to inferior; ovary 2 locular or 3-4 locular. Styles usually 1 or rarely 2-4; free (rarely) or partially joined; apical placentation. Ovules 2 per locule, pendulous, anatropous ovule

→ Fruit rarely fleshy or non-fleshy; indehiscent; a drupe

→ Seeds non endospermic with a straight well differentiated embryo.

Genera included under chailletiaceae

    → Dichapetalum
    → Stephanopodium
    → Tapura

Sub class: Polypetalae

Series: Discinflorae

Order: Olacales

Families:

Olacineae

Ilicineae

Cyrilleae

Chapter-47
Olacineae (Olacaceae)
(Bentham and Hooker)

Classification

Phanerogams
Dicotyledons
Polypetalae
Disciflorae
Olacales
Olacineae

General Characters

→ Trees, shrubs and lianas
→ Plants - autotrophic or parasitic, when parasitic haustoria present.
→ Leaves - alternate, distichous or rarely spiral, herbaceous to leathery, petiolate, or subsessile to sessile, non-sheathing, gland dotted or not gland dotted, simple, lamina entire, one veined or pinnately veined, exstipulate, margin entire
→ Inflorescence - Solitary (olax), aggregated in inflorescence - Panicles, racemes or heads. When solitary, axillary.
→ Flowers - Plants hermaphrodite or monoecious or dioecious; small, regular, cyclic. Hypogynous disc present, annular
→ Perianth with distinct calyx and corolla. Calyx consists of 3-6 sepals in 1 whorl, gamosepalous

entire or lobulate or blunt-lobed or toothed. Cupuliform; regular, fleshy or non-fleshy, persisten often accrescent; imbricate

→ Corolla consists of 3-6 petals in 1 whorl. Polypetalous or gamopetalous; Valvate, regular, non fleshy

→ Androecium consists of 3-6, 6-12 or 9-18 stamens Stamens - polystemonous or number relative to that of perianth; alternisepalous or oppositisepalous Anthers basifixed, versatile or non-versatile or non-versatile, dehisce via pores or longitudinal slits.

→ Gynoecium - 2-5 carpelled. Carpels isomerous with the perianth or reduced in number relative to that of perianth. Syncarpous, superior to partly inferior, ovary - 2-5 locular or rarely 1 locular. Styles 1, apical; Stigmas 2-5 lobed. Placentation when unilocular is free central, others - axile to apical placentation. Ovules 1 per locule, pendulous, anatropous

→ Fruit fleshy, or non fleshy, indehiscent, a drupe or a nut; enclosed in a fleshy hypanthium or in the fleshy perianth

→ Seeds endospermic with straight embryo

Genera included under Olacaceae

- Anacolosa
- Aptandra
- Brachynema
- Cathedra
- Chaunochiton
- Coula
- Curupira
- Diogoa
- Douradoa
- Dulacia
- Harmandia
- Heisteria
- Malania
- Minquartia
- Ochanostachys
- Olax
- Ongokea
- Phanerodiscus
- Ptychopetalum
- Schoepfia
- Scorodocarpus
- Strombosia
- Strombosiopsis
- Tetrastylidium
- Ximenia.

Chapter-48
Ilicineae (Aquifoliaceae)

Classification (Bentham and Hooker)

Phanerogams
Dicotyledons
Polypetalae
Disciflorae
Olacales
Ilicineae

General characters

→ Trees and Shrubs
→ Ovary entire, 1-α celled, Stamens hypogynous or subhypogynous. Ovules 1-3 in each cell, Pendulous, with a dorsal raphe
→ Seeds usually albuminous.

Genera included under Ilicineae

→ Ilex
→ Nemopanthus

Chapter - 49
Cyrilleae

Classification (Bentham and Hooker)

- Phanerogams
- Dicotyledons
- Polypetalae
- Disciflorae
- Olacales
- Cyrilleae

General characters

→ Trees and shrubs
→ Ovary entire, 1-2 celled, stamens hypogynous or subhypogynous. Ovules 1-3 in each cell, pendulous, with a dorsal raphe
→ Seeds usually albuminous.

Sub class: polypetalae

Series: Disciflorae

Order: Celastrales

Families:

Celastrineae

Stackhousieae

Rhamneae

Ampelidea

Chapter - 50
Celastrineae (celastraceae)

Classification (Bentham and Hooker)

- Phanerogams
- Dicotyledons
- Polypetalae
- Disciflorae
- Celestrales
- Celastrineae

General characters

→ Trees and shrubs; sometimes climbing or vining
→ Leaves - simple, alternate or opposite, stipulate, stipules small and caducous or absent
→ Flowers - bisexual or sometimes functionally unisexual, actinomorphic and are small often greenish.
→ Calyx consists of 4 or 5 basally connate sepals.
→ Corolla is rarely absent or usually consists of 4 or 5 distinct petals.
→ Androecium consists of 4 or 5 or rarely 10 distinct stamens which alternate the petals
→ Gynoecium is a single compound pistil of 2-5 carpels, a single short style and a superior or half inferior ovary with 2-5 locules, each

containing usually 2 axile ovules.
- → An annular nectary disk surrounds and is usually adnate to the ovary.
- → Fruit - capsule, berry, samara, drupe

Genera included under Celastraceae

- → Acanthothamnus
- → Allocassine
- → Anthodon
- → Apatophyllum
- → Apodostigma
- → Arnicratea
- → Bequaertia
- → Brassiantha
- → Brexia
- → Brexiella
- → Campylostemon
- → Canotia
- → Cassine
- → Catha
- → Celastrus
- → Cheiloclinium
- → Crocoxylon
- → Crossopetalum
- → Cuervea
- → Denhamia
- → Dicarpellum
- → Dinghoua
- → Elachyptera
- → Elaeodendron
- → Empleuridium
- → Euonymus
- → Euonymopsis
- → Fraunhofera
- → Gloveria
- → Glyptopetalum
- → Goniodiscus
- → Gyminda
- → Gymnosporia
- → Hartogiella
- → Hartogiopsis
- → Hedraianthera
- → Helictonema
- → Hexaspora
- → Hippocratea
- → Hylenaea
- → Hypsophila
- → Kokoona
- → Lauridia
- → Loeseneriella
- → Lophopetalum
- → Lydenburgia
- → Macgregoria
- → Maurocenia
- → Maytenus
- → Menepetalum
- → Microtropis
- → Monimopetalum
- → Mortonia
- → Moya
- → Mystroxylon
- → Orthosphenia
- → Parnassia
- → Paxistima
- → Peripterygia
- → Peritassa
- → Plagiopteron
- → Platypterocarpus
- → Plenckia
- → Pleurostylia
- → Polycardia

- Pleonostemma
- Prutimera
- Psammomoya
- Pseudocatha
- Pseudosalacia
- Pteledeum
- Pterocelastrus
- Putterlickia
- Quetzalia
- Reissantia
- Robsonodendron
- Rzedowskia
- Salacia
- Salacighia
- Salaciopsis
- Sarawakodendron
- Scandivepres
- Schaefferia
- Semialarium
- Simicratea
- Simirestis
- Siphonodon
- Stackhousia
- Tetrasiphon
- Thyrsosalacia
- Tontelea
- Torralbasia
- Tricerma
- Tripterococcus
- Tripterygium
- Tristemonanthus
- Wimmeria
- Xylonymus
- Zinowiewia

Fossil genera
- Celastrinites

Chapter-51
Stackhousieae

Classification (Benthary and Hooker)

- Phanerogams
- Dicotyledons
- Polypetalae
- Disciflorae
- Celastrales
- Stackhousieae

⇒ Now merged into celestraceae family. When accepted, it comprised the following genera
- → Macgregoria
- → Stackhousia
- → Tripterococcus

Chapter-52
Rhamneae (Rhamnaceae)

Classification (Bentham and Hooker)

Phanerogams
Dicotyledons
Polypetalae
Disciflorae
Celastrales
Rhamneae

General characters

→ Plants – trees, shrubs or climbers, climbing by hooks or tendrils;
→ Leaves – single, stipulate, stipules often spiny
→ Inflorescence – cymose
→ Flowers – hermaphrodite, perigynous, sepals 4-5, free; petals 4-5, free; stamens opposite the concave petals; a well developed intrastaminal disc present
→ Carpels 2-4 and with 2-4 locules; one basal ovule in each locule
→ Fruit is a drupe or capsule
→ Seeds hard

Genera included under Rhamnaceae

- Ampelozezyphus
- Bathiorhamnus
- Adolphia
- Colletia
- Discaria
- Kentrothamnus
- Ochetophila
- Retanilla
- Trevoa
- Doerpfeldia
- Alvimiantha
- Crumenaria
- Gouania
- Helinus
- Johnstonalia
- Pleuranthodes
- Reissekia
- Maesopsis
- Hovenia
- Paliurus
- Ziziphus
- Sarcomphalus
- Nesiota
- Noltea
- Phylica
- Trichocephalus
- Blackallia
- Cryptandra
- Papistylus
- Polianthion
- Pomaderris
- Serichonus
- Siegfriedia
- Spyridium
- Stenanthemum
- Trymalium
- Auerodendron
- Berchemia
- Berchemiella
- Condalia
- Frangula
- Karwinskia
- Krugiodendron
- Reynosia
- Rhamnella
- Rhamnidium
- Rhamnus
- Sageretia
- Scutia
- Smythea
- Ventilago
- Alphitonia
- Araracuara
- Ceanothus
- Chaydaia
- Colubrina
- Emmenosperma
- Granitites
- Hybosperma
- Jaffrea
- Lasiodiscus
- Schistocarpaea
- Talguenea

Chapter - 53
Ampelidaceae (Ampelideae)

Classification (Bentham and Hooker)

Phanerogams
Dicotyledons
Polypetalae
Disciflorae
Celastrales
Ampelidaceae

General characters

→ Climbing shrubs or small trees.
→ Leaves - compound or simple deeply lobed
→ Flowers - small, hermaphrodite or polygamous-dioecious in spikes, racemose, panicles or cymes
→ Sepals 4-5, connate, cup shaped
→ Petals 4-5, polypetalous or connate at the apex, caducous;
→ Stamens 4-5, antipetalous
→ Carpels 2-8, syncarpous, ovary superior
→ Fruit - juicy berry.

Genera included under Ampelidaceae

→ Acareosperma
→ Ampelocissus
→ Ampelopsis

- Cayratia
- Cissus
- Clematicissus
- Cyphostemma
- Leea
- Nekemias
- Nothocissus
- Parthenocissus
- Pterisanthes
- Pterocissus
- Rhoicissus
- Tetrastigma
- Vitis
- Yua.

Sub class: polypetalae

Series: Disciflorae

Order: Sapindales

Families:

Sapindaceae

Sabiaceae

Anacardiaceae

Coriarieae

Moringeae

Chapter - 54
Sapindaceae

Classification (Bentham and Hooker)

Phanerogams
Dicotyledons
Polypetalae
Disciflorae
Sapindales
Sapindaceae

General characters

→ Trees, shrubs or climbers by tendrils.
→ Anamolous secondary thickening takes place as a result of unusual position of primary vascular bundles in many climbing genera.
→ Leaves - alternate, rarely opposite, usually compound, pinnate, sometimes simple (*Cardiospermum*) exstipulate or stipules sometimes present in climbing species. (*Melianthus*)
→ Inflorescence - cymose or racemose. Small flowers are found to be arranged in racemose inflorescence or in paniculate cymes.
→ Flower - Obliquely zygomorphic or actinomorphic, hermaphrodite or unisexual, hypogynous, polygamous or polygamodioecious, pentamerous rarely tetramerous.

unilateral, annular or glandular extrastaminal disc present. The flowers may be bracteate or bracteolate.
→ Calyx consists of 5 sepals, free, valvate or imbricate aestivation, in actinomorphic flowers sepals 4 by the fusion of 3rd and 5th sepals.
→ Corolla – petals 4 or 5, free, provided with a scaly or hairy appendages. In actinomorphic condition, the pentamerous corolla become tetramerous by the suppression of 1 petal. Petals absent in some (*Dodonaea*)
→ Androecium – usually 8 or 10 in 2 whorls, free sometimes eccentric. Filaments hairy, free or basally connate (*Cardiospermum*). Anthers – dithecous, basifixed and introrse.
→ Gynoecium – Tricarpellary, Syncarpous, ovary superior, trilocular, one or two ovules in each locule, style terminal, axile placentation. The characteristic of this family is the constant number of ovules (1 or 2)
→ Fruit – capsule, nut, drupe, berry. In *Acer* the fruit is double samara.
→ Seed – arillate. In *Litchi chinensis*, the aril is pulpy and juicy forming the edible part of the fruit. Endospermic or non endospermic with plicate or convolute embryo. Pollination – entomophilous.

Genera included under Sapindaceae

- Alectryon
- Allophylus
- Allosanthus
- Amesiodendron
- Aporrhiza
- Arfeuillea
- Arytera
- Atalaya
- Athyana
- Averrhoidium
- Beguea
- Bizonula
- Blighia
- Blighiopsis
- Blomia
- Boniodendron
- Bridgesia
- Camptolepis
- Cardiospermum
- Castanospora
- Chonopetalum
- Chouxia
- Chytranthus
- Conchopetalum
- Cossinia
- Cubilia
- Cupania
- Cupaniopsis
- Deinbollia
- Delavaya
- Diatenopteryx
- Dictyoneura
- Dilodendron
- Dimocarpus
- Diploglottis
- Diplokelepa
- Diplopeltis
- Distichostemon
- Dodonaea
- Doratoxylon
- Elattostachys
- Eriocoelum
- Erythrophysa
- Euchorium
- Euphorianthus
- Eurycorymbus
- Exotheca
- Filicium
- Ganophyllum
- Glennea
- Gleocarpus
- Gongrodiscus
- Gongrospermum
- Guindilia
- Guioa
- Handeliodendron
- Haplocoelum
- Harpullia
- Hippobromus
- Hornea
- Houssayanthus
- Hypelate
- Hypselodernia
- Jagera
- Koelreuteria
- Laccodiscus
- Lecaniodiscus
- Lepiderema
- Lepidopetalum
- Lepisanthes
- Litchi
- Llagunoa
- Lophostigma
- Loxodiscus
- Lychnodiscus

Families of Dicotyledons

- Macphersonia
- Magonia
- Majidea
- Matayba
- Melicoccus
- Mischocarpus
- Molinaea
- Neotina
- Nephelium
- Otonephelium
- Pancovia
- Pappea
- Paranephelium
- Paullinia
- Pavieasia
- Pentascyphus
- Phyllotrichum
- Placodiscus
- Plagioscyphus
- Podonephelium
- Pometia
- Porocystis
- Pseudima
- Pseudopancovia
- Pseudopteris
- Radlkofera
- Physotoechia
- Sapindus
- Sarcopteryx
- Sarcotoechia
- Schleichera
- Scyphonychium
- Serjania
- Seshoradlkofera
- Sesyrolepis
- Smelophyllum
- Stadmania
- Stocksia
- Storthocalyx
- Synima
- Talisia
- Thinouia
- Thouinidium
- Tina
- Tinopsis
- Toechima
- Toulicia
- Trigonachras
- Tripterodendron
- Tristira
- Tristiropsis
- Tsingya
- Ungnadia
- Urvillea
- Vouarana
- Xanthoceros
- Xerospermum
- Zanha
- Zollingera

Chapter - 55
Sabiaceae

Classification: (Bentham and Hooker)

- Phanerogams
- Dicotyledons
- Polypetalae
- Disciflorae
- Sapindales
- Sabiaceae

General characters

→ Trees, climbing shrubs or woody vines
→ Leaves - alternate, spiral to distichous, pennenerved, simple, imparipinnate, herbaceous to coriaceous, often the base of the stalk is woody and the base of the leaf is pulvinulate, exstipulate.
→ Plants hermaphrodite rarely polygamodioecious
→ Inflorescence - Panicle, terminal or axillary, often reduced to solitary axillary, rarely on cymes or racemes
→ Flower - actinomorphic or obliquely zygomorphic, pentamerous, sepals, petals and stamens are arranged in opposite whorls. Hypogynous disc present.
→ Calyx consists of 4-5 sepals, free or basally connate and 1 whorled, Imbricate aestivation

- → Corolla consists of 4-5 petals, 1 whorled, imbricate oppositisepalous, more or less fleshy
- → Androecium consists of 4-6 stamens or even 2 and 3 staminodes, oppositipetalous, free from each other but fused at the base of the petals, filaments filiform, expanded below the anther or forming a collar, unilocular anthers, dithecal, introrse and bent down enclosed in external cavities belonging to adjacent staminode.
- → Gynoecium - 2-3 carpels, superior ovary, styles free or sincarpic with one short cylindrical or conic style, capitate stigmas. Ovules 1-2 per carpel, hemianatropous to campylotropous, axile placentation
- → Fruit - unilocular or bilocular, asymmetric, dry or drupaceous, indehiscent, sometimes schizocarp with persistent styles.
- → Seed - 1, endosperm scarce or absent, embryo curved.

Genera included under Sabiaceae

- → Sabia

Chapter-56
Anacardiaceae

Classification (Bentham and Hooker)

Phanerogams
Dicotyledons
Polypetalae
Disciflorae
Sapendales
Anacardiaceae

General Characters

→ Mainly trees or shrubs. Resin canals are present in the wood.
→ Branched Tap root system
→ Stem - erect, branched, woody, cylindrical, solid
→ Leaves - simple or compound; alternate & petiolate, exstipulate; Leaves rarely opposite (Dobinea)
→ Flower - hermaphrodite, but sometimes unisexual by the reduction of any of the sexes. Usually actinomorphic and Pentamerous. They may be hypo, epi or perigynous.
→ Calyx consists of 3 to 5 sepals, gamosepalous. Sometimes, the sepals are adnate to the ovary wall.
→ Corolla consists of 3 to 7 petals, Polypetalous or some cases united at the base. Rarely corolla altogether absent

→ Androecium consists of 10 stamens arranged in two whorls of 5 each. Very often the number of stamens double that of petals. Stamens may be free or united at the base. They arise from the edge of an intrastaminal disc. Anthers are dithecous. Split by longitudinal slits.

→ Gynoecium consists of 3 carpels, syncarpous. Out of the 3 carpels only one is functional. Ovary superior rarely inferior and contains a single, solitary, pendulous ovule. In most of the cases only one ovule matures and becomes seed. Placentation — axile. Each locule contains a single ovule. Style — short; Stigmas are as many as the number of carpels.

→ Fruit — usually a drupe

→ Seeds — non-endospermic and possesses a very little endosperm. Embryo bears fleshy cotyledons.

→ Pollination — entomophilous.

Genera included under Anacardiaceae

- Actinocheita
- Anacardium
- Androtium
- Antrocaryon
- Apterokarpos
- Astronium
- Baronia
- Bonetiella
- Bouea
- Buchanania
- Campnosperma
- Cardenasiodendron
- Choerospondias
- Comocladia
- Cotinus
- Cyrtocarpa
- Dracontomelon
- Drimycarpus
- Ebandoua
- Euleria
- Euroschinus
- Faguetia
- Fegimanra
- Gluta
- Haematostaphis
- Haplorhus
- Harpephyllum
- Heeria
- Holigarna
- Koordersiodendron
- Lannea
- Laurophyllus
- Lithrea
- Loxopterigium
- Loxostylis
- Mangifera
- Mauria
- Melanochyla
- Metopium
- Micronychia
- Montagueia
- Mosquitoxylum
- Nothopegia
- Ochoterenaea
- Operculicarya
- Ozoroa
- Pachycormus
- Pareishia
- Pegia
- Pentaspadon
- Pleiogynium
- Poupartia
- Protorhus
- Pseudoprotorhus
- Pseudosmodingium
- Pseudospondias
- Rhodosphaera
- Rhus
- Schinopsis
- Schinus
- Sclerocarya
- Searsia
- Semecarpus
- Smodingium
- Solenocarpus
- Sorindeia
- Spondias
- Swintonia
- Tapirira
- Thyrsodium
- Toxicodendron
- Trichoscypha

Chapter - 57
Coriarieae (Coriariaceae)

Classification (Bentham and Hooker)

- Phanerogams
- Dicotyledons
- Polypetalae
- Disciflorae
- Sapindales
- Coriarieae

General characters

→ Deciduous shrub
→ Mostly shrubs or subshrubs or small trees
→ Leaves - simple, opposite or whorled, with 3 to 9 major veins coming from base
→ Fruit - look like berry but they are small nuts (achenes) protected by enlarged and coloured petals.
→ Seeds poisonous
→ Several species can fix nitrogen from the air because they have bacteria of genus *Frankia* in their roots.

Genera belonging to Coriarieae

→ Coriaria

Chapter-58
Moringaceae

Classification (Bentham and Hooker)

Phanerogams
Dicotyledons
Polypetalae
Disciflorae
Sapindales
Moringaceae

General Characters

→ Trees, deciduous with gummy bark
→ Tap, branched and deep root
→ Stem - erect, branched, woody, wood fragile
→ Leaf - Pinnately compound with opposite pinnae, alternate, stipules are reduced to glands or absent, pulvinus of petioles distinct.
→ Inflorescence - Hairy axillary cymose panicles
→ Flower - Hermaphrodite, Medianly zygomorphic or actinomorphic, perigynous, more or less complete, pedicellate, bracteate
→ Calyx consists of 5 sepals, polysepalous, imbricate, reflexed or spreading, inserted on the margin of cupular receptacle - hypanthium.
→ Corolla consists of 5 petals, polypetalous, inserted on hypanthium rim, the 2 posterior

ones smaller and reflexed with 2, erect laterals ascending and the anterior ones larger, imbricate, disc lining the hypantheum is present
→ Androecium consists of 5 stamens, fertile, polyandrous, filaments unequal in length, inserted on the margin of the disc, staminodes 3-5, alternating with fertile stamens, anthers monothecal, dehisce by longitudinal slits.
→ Gynoecium - 3 carpels, syncarpous, unilocular, superior ovary, style one, long, curved terminate into flat or truncate or club shaped stigma. Ovules many, biseriate, anatropous on parietal placentation
→ Fruit - capsule
→ Seed - endospermic, seeds may or may not be winged, cotyledons oil
→ Pollination entomophilous

Genera included under Moringaceae
→ Moringa

Sub class: Polypetalae

Series: Calyciflorae

Order: Rosales

Families:

Connaraceae

Leguminosae

Rosaceae

Saxifrageae

Crassulaceae

Droseraceae

Hamamelideae

Bruniaceae

Halorageae

Chapter - 59
Connaraceae

Classification (Bentham and Hooker)

Phanerogams
Dicotyledons
Polypetalae
Calyciflorae
Rosales
Connaraceae

General Characters

→ Trees, shrubs and shrubby, twining climbers
→ Leaves- alternate, spiral, leathery, Petiolate, non-sheating; compound; exstipulate
→ Plants usually hermaphrodite or dioecious.
→ Flowers aggregated in inflorescence - racemes and Panicles. Flowers small, regular or somewhat irregular, Pentamerous.
→ Perianth with distinct calyx and corolla. 8-10 in 2 whorl, isomerous. Calyx 4-5 in 1 whorl; Poly or gamosepalous; persistent; imbricate or valvate
→ Corolla 4-5 in 1 whorl; Polypetalous or gamopetalous usually imbricate or rarely valvate, regular.

→ Androecium usually 10 or rarely 5; free of the perianth; usually free of one another or rarely connate at the base; usually 2 whorled.
→ Stamens 5-10
→ Anthers dorsifixed, versatile, dehiscing via longitudinal slits; introrse.
→ Gynoecium - 1 carpelled or 3 carpelled rarely 5, 7 or 8 carpelled; Placentation - Marginal or basal. Gynoecium monomerous or apocarpous. Carpel apically stigmatic
→ Fruit - non fleshy often a single follicle rarely aggregate.
→ Seeds endospermic or non-endospermic. Endosperm when present, oily.

Genera included under Connaraceae

→ Agelaea
→ Burttia
→ Cnestidium
→ Cnestis
→ Connarus
→ Ellipanthus
→ Hemandradenia
→ Jollydora
→ Manotes
→ Pseudoconnarus
→ Rourea
→ Roureopsis
→ Taeniochlaena
→ Vismianthus

Chapter 60: Leguminosae

Caesalpiniaceae

Classification (Bentham and Hooker)

- Phanerogams
- Dicotyledons
- Polypetalae
- Calyciflorae
- Rosales
- Caesalpiniaceae

General Characters:

→ Trees or shrubs, rarely herbs

→ Mostly Mesophytes, but Xerophytes are also reported (Parkinsonia)

→ Branched Tap root system

→ Stem - erect, woody, cylindrical, solid, branched sometimes herbaceous or climbing. Tannin sacs and gum passages are found in many species

→ Leaves - Simple or compound. If compound - pinnate or bipinnate; usually pinnate leaves are arranged in pairs; Petiolate, Pulvinus present at the base of the petiole; Pinna - ovate or obovate, glabrous, net veined, entire; usually exstipulate, sometimes minute Caducous Stipules present

→ Inflorescence - usually racemose, raceme sometime pendulous.

- → Flowers – Pedicellate, zygomorphic, rarely actinomorphic, hermaphrodite, hypogynous, complete, variously coloured, showy, large or small
- → Calyx consists of 5 sepals, free or fused, often petaloid; Imbricate or valvate aestivation.
- → Corolla consists of 5 petals, free (polypetalous); ascending imbricate aestivation; inferior; Spathulate, showy.
- → Androecium – 10 stamens, free or rarely connate but sometimes reduced to staminodes.
- → Gynoecium – Monocarpellary (1 carpel); ovary superior; unilocular, marginal placentation; style long, simple stigma.
- → Fruit – Pod or loment, sessile or stalked, dehiscent or indehiscent
- → Exalbuminous seed.

Genera included under Caesalpiniaceae

- → Caesalpinia
- → Acrocarpus
- → Bauhinia
- → Brachystegia
- → Cassia
- → Ceratonia
- → Cercidium
- → Chamaecrista
- → Delonix
- → Gleditsia
- → Gymnocladus
- → Haematoxylon
- → Parkinsonia
- → Petteria
- → Senna
- → Saraca
- → Tamarindus
- → Poinciana
- → Libidibia
- → Lysiphyllum
- → Maniltoa
- → Peltophorum

Genera included under Caesalpiniaceae

- Acrocarpus
- Adenolobus
- Afzelia
- Amherstia
- Androcalymma
- Anthonotha
- Apaloxylon
- Aphanocalyx
- Aprevalia
- Apuleia
- Arapatiella
- Arcoa
- Augouardia
- Baikiaea
- Barklya
- Batesia
- Bathiaea
- Baudouinia
- Bauhinia
- Berlinia
- Brachycylix
- Brachystegia
- Bracteolanthus
- Brandzeeria
- Brodriguesia
- Brownea
- Browneopsis
- Burkea
- Bussea
- Caesalpinia
- Campsiandra
- Candolleodendron
- Cassia
- Cenostigma
- Ceratonia
- Cercidium
- Cercis
- Chidlowia
- Colophospermum
- Colvillea
- Conzattia
- Copaifera
- Cordeauxia
- Crudia
- Cryptosepalum
- Cymometra
- Daniella
- Dansera
- Delonix
- Detarium
- Dialium
- Dicorynia
- Dicymbe
- Didelotia
- Dimorphandra
- Diptychandra
- Distemonanthus
- Duparquetia
- Elegmocarpus
- Elizabetha
- Endertia
- Englerodendron
- Eperua
- Erythrophleum
- Eurypetalum
- Gigasiphon
- Gilbertiodendron
- Gilletiodendron
- Gleditsia
- Gonorrachis
- Gossweilerodendron
- Graeffonia
- Guibourtia
- Gymnocladus
- Haematoxylum
- Hardwickia
- Heterostemon

- Hoffmannseggia
- Holocalyx
- Humboldtia
- Hylodendron
- Hymenaea
- Hymenostegia
- Intsia
- Isoberlinia
- Jacqueshuberia
- Julbernardia
- Kalappia
- Kaoue
- Kingiodendron
- Koompassia
- Labichea
- Lasiobema
- Lebruniodendron
- Lemuropsium
- Leonardoxa
- Leucostegane
- Moldenhauera
- Monopetalanthus
- Mora
- Neochevalierodendron
- Oddoniodendron
- Orphanodendron
- Oxystigma
- Pachyelasma
- Paloue
- Paloveopsis
- Paramacrolobium
- Parkinsonia
- Pellegriniodendron
- Peltogyne
- Peltophorum
- Petalostyles
- Phanera
- Phyllocarpus
- Pileostigma
- Plagiosiphon
- Poeppigia
- Polystemonanthus
- Prioria
- Pseudomacrolobium
- Pterogyne
- Pterolobium
- Recordoxylon
- Saraca
- Schizolobium
- Schizocyphus
- Schotia
- Sclerolobium

- Scorodophloeus
- Sindora
- Sindoropsis
- Stachyothyrsus
- Stahlia
- Stemonocoleus
- Stenodrepanum
- Storckiella
- Stuhlmannia
- Sympetalandra
- Tachigalia
- Talbotiella
- Tamarindus
- Tessmannia
- Tetraberlinia
- Tetrapterocarpon
- Thylacanthus
- Trachylobium
- Uittenia
- Uritiza
- Vouacapoua
- Wagatea
- Zenia
- Zenkerella
- Zuccagnia.

Mimosaceae

Classification (Bentham and Hooker)

Phanerogams
Dicotyledons
Polypetalae
Calyciflorae
Rosales
Leguminosae
Mimosaceae

General Characters:

→ Either shrubs or trees, very rarely herbs, sometimes climbers.

→ Most are thorny and xeromorphic; some hydrophytes are also found (Neptunia)

→ Branched tap root, deep rooted in the soil

→ Stem - erect, branched, terete, woody, solid. Stem tissues is often rich in tannin sacs and gum passages.

→ Leaves - Alternate, petiolate, usually the base of petiole is provided with pulvinus; stipulate, usually stipules are modified into thorns; compound, pinnate, generally bipinnate.

→ Inflorescence - Racemose, head or spike

→ Flower - usually sessile, actinomorphic, regular, hermaphrodite, hypogynous, complete, small.

→ Calyx consists of 5 or 4 sepals, gamosepalous, the sepals more or less connate, green (sepaloid), small, inferior, valvate aestivation

→ Corolla consists of 5 or 4, free petals (polypetalous) valvate; inferior, slightly united towards the base, usually pentamerous

→ Androecium – usually indefinite stamens, but sometimes reduced to 10 or even 4 (*Mimosa pudica*) Stamens are conspicuous, attractive, bright coloured and somewhat scented. Anthers 2 celled, dorsifixed, dehisce by longitudinal slits. The filaments are long and slender

→ Gynoecium consist of one carpel (monocarpellary); Ovary – superior, unilocular; Placentation – marginal; Style long filiform; stigma – terminal and simple

→ Fruit – legume or lomentum
→ Seeds – Exalbuminous
→ Pollination is entomophilous.

Genera included under Mimosaceae

- Abarema
- Acacia
- Albizia
- Adenanthera
- Adenopodia
- Affonsea
- Albizia
- Amblygonocarpus
- Adenanthera
- Archidendron
- Archidendropsis
- Aubrevillea
- Calliandra
- Calliandropsis
- Calpocalyx
- Cedrelinga
- Copoba
- Cylicodiscus
- Desmanthus
- Dichrostachys
- Elephantorrhiza
- Entada
- Enterolobium
- Faidherbia
- Fillaeopsis
- Gagnebina
- Goldmania
- Havardia
- Indopiptadenia
- Lemurodendron
- Leucaena
- Lysiloma
- Marmaroxylon
- Mimosa
- Mimozyganthus
- Neptunia
- Newtonia
- Parapiptadenia
- Parachidendron
- Paraserianthes
- Parkia
- Pentaclethra
- Piptadenia
- Piptadeniastrum
- Piptadeniopsis
- Pithecellobium
- Plathymenia
- Prosopidastrum
- Prosopis
- Pseudoentada
- Pseudopiptadenia
- Pseudoprosopis
- Sclerinitzia
- Serianthes
- Stryphnodendron
- Tetrapleura
- Wallaceodendron
- Xerocladia
- Xylia
- Zapoteca
- Zygia

Fabaceae (Papilionaceae)

Classification (Bentham and Hooker)

- Phanerogams
- Dicotyledons
- Polypetalae
- Calyciflorae
- Leguminosae
- Fabaceae

General characters:

→ Mostly herbs or shrubs, often climbing; very rarely trees.

→ Root is tap, branched, bearing nodules filled with <u>Rhizobia</u> (Nitrogen fixing bacterium) and hence the plants are used as green manures.

→ Stem - erect, climbing (with the aid of tendrils), branched, angular or cylindrical, herbaceous or woody.

→ Leaves - alternate, opposite or whorled, usually compound (digitate or pinnate), rarely even pinnate, sometimes simple, stipulate, stipules occur at the base of the petiole (foliaceous and large). In some cases, secondary stipules arise at the base of the individual leaflets.

→ Inflorescence - racemose, raceme, spike and contracted raceme or head.

→ Flower – Pedicellate, zygomorphic, irregular, hermaphrodite, complete, Perigynous and Papilionaceous.

→ Calyx consists of united sepals (gamosepalous), equal or unequal, below the disc united in a tubular calyx, 5 toothed, or 5 lobed or bilabiate (2 upper and 3 lower may unite). Ascending imbricate aestivation.

→ Corolla consists of 5 petals (unequal), uppermost large petal is known as standard or vexillum; the two free lateral petals are called wings or alae; anterior pair of united petals is called keel or carina which encloses stamens and pistil. Descending imbricate aestivation.

→ Androecium – Stamens are usually 10, inserted on a disc below the calyx, they may be in 2 bundles (diadelphous) of 9+1 or 5+5, or in one bundle (monadelphous), rarely free. Anthers bi-fixed, dorsifixed, dehisce by longitudinal slits.

→ Gynoecium – Carpel one, free, superior ovary, stalked or sessile, unilocular, marginal placentation, style bent and short, hairy, stigma simple.

→ Fruit – legume or pod, splitting along both dorsal and ventral sutures. It is indehiscent lomentum in some species.

→ Seeds – many, exalbuminous, usually flattened.

Families of Dicotyledons

Genera included under Papilionaceae

- Abrus
- Acmespon
- Acosmium
- Adenocarpus
- Adenodolichos
- Adesmia
- Aenictophyton
- Aeschynomene
- Afgekia
- Aganope
- Airyantha
- Aldina
- Alexa
- Alhagi
- Alistelus
- Almaleea
- Alysicarpus
- Amburana
- Amecia
- Ammodendron
- Ammopiptanthus
- Ammothamnus
- Amphiodon
- Amorpha
- Amphicarpaea
- Amphimas
- Amphithalea
- Anagyris
- Anarthrophyllum
- Ancistrotropis
- Andira
- Angylocalyx
- Antheroporum
- Anthyllis
- Antopetitia
- Aotus
- Aphyllodium
- Apios
- Apoplanesia
- Apurimacia
- Arachis
- Argyrocytisus
- Argyrolobium
- Arthrocleanthus
- Aspalathus
- Astragalus
- Ateleia
- Austrodolichos
- Austrosteenesia
- Baphia
- Baphiastrum
- Baphiopsis
- Baptisia
- Barbieria
- Behaimia
- Bionia
- Bituminaria
- Bolgunnia
- Bocoa
- Bolusafra
- Bolusanthus
- Bolusia
- Bossiae
- Bowdichia
- Bowringia
- Brongniartia
- Brya
- Bryaspis
- Burkilliodendron
- Butea
- Cadia
- Cajanus
- Calia
- Calicotome
- Callerya
- Callistachys

- Dipteryx
- Discolobium
- Duynstenion
- Dolichopsis
- Dolichos
- Dorycnium
- Droogmansia
- Dumasia
- Dunbaria
- Dussia
- Dysolobium
- Ebenus
- Echinospartum
- Eleiotis
- Eminia
- Endosamara
- Eremosparton
- Erichsenia
- Erinacea
- Errosema
- Errazurezia
- Erythrina
- Etaballia
- Euchilopsis
- Euchlora
- Euchresta
- Eutaxia
- Eversmannia
- Exostyles
- Eysenhardtia
- Ezoloba
- Fairchildia
- Flemingrella
- Fissicalyx
- Flemingia
- Fordia
- Galactia
- Galega
- Gastrolobium
- Geissaspis
- Genista
- Genistidium
- Geoffroea
- Gliricidia
- Glycine
- Glycyrrhiza
- Gompholobium
- Gonocytisus
- Goodia
- Grazielodendron
- Guianodendron
- Gueldenstaedtia
- Halimodendron
- Hammatolobium
- Haplormosia
- Hardenbergia
- Harleyodendron
- Harpalyce
- Hebestigma
- Hedysarum
- Heliotropis
- Herpyza
- Hesperolaburnum
- Hippocrepis
- Hoita
- Holocalyx
- Hosackia
- Hovea
- Humularia
- Hymenocarpus
- Hymenolobium
- Hypocalyptus
- Indigastrum
- Indigofera
- Inocarpus
- Isotropis
- Jacksonia
- Kennedia
- Kotschya

- Calobota
- Calophaca
- Calopogonium
- Calpurnia
- Camoensia
- Camptosema
- Campylotropis
- Canavalia
- Candolleodendron
- Caragana
- Carmichaelia
- Carrissoa
- Cascaronia
- Castonospermum
- Centrolobium
- Centrosema
- Chadsia
- Chaetocalyx
- Chamaecytisus
- Chapmannia
- Chesneya
- Chorizema
- Christia
- Cicer
- Cladrastis
- Clathrotropis
- Cleobulia
- Cleanthus
- Clitoria
- Clitoriopsis
- Cochlianthus
- Cochleasanthus
- Codariocalyx
- Collaea
- Cologania
- Colutea
- Condylostylis
- Cordyla
- Coronilla
- Coursetia
- Craibia
- Cranocarpus
- Craspedolobium
- Cratylia
- Cristonia
- Crotalaria
- Cruddasia
- Cullen
- Cyamopsis
- Cyathostegia
- Cyclocarpa
- Cyclolobium
- Cyclopia
- Cymbosema
- Cyclopia
- Cytisopsis
- Cytisus
- Dahlstedtia
- Dalbergia
- Dalbergiella
- Dalea
- Dalhousiea
- Daviesia
- Decorsea
- Dendrolobium
- Derris
- Dermatophyllum
- Desmodiastrum
- Desmodium
- Dewevrea
- Dichilus
- Dicraeopetalum
- Dillwynia
- Dioclea
- Diphyllarium
- Diphysa
- Diplotropis
- Dipogon

- Kummerowia
- Lablab
- Laburocytisus
- Laburnum
- Lackeya
- Lamprolobium
- Lathyrus
- Latrobea
- Lebeckia
- Lecointea
- Lembotropis
- Lennea
- Lens
- Leobordea
- Leptoderris
- Leptodesmia
- Leptolobium
- Leptosema
- Leptospron
- Lespedeza
- Lessertia
- Leucomphalos
- Limadendron
- Liparia
- Listia
- Lonchocarpus
- Lotononis
- Lotus
- Luetzelburgia
- Lupinus
- Luzonia
- Maackia
- Machaerium
- Macropsychanthus
- Macroptilium
- Macrotyloma
- Maraniona
- Margaritolobium
- Marina
- Mastersia
- Mecopus
- Medicago
- Melilotus
- Mellinsella
- Melolobium
- Microcharis
- Mildbraediodendron
- Millettia
- Mirbelia
- Monopteryx
- Mucuna
- Muellera
- Muelleranthus
- Mundulea
- Myrocarpus
- Mysospermum
- Myroxylon
- Mysanthus
- Neocollettia
- Neoharmsia
- Neonotonia
- Neorautanenia
- Neorudolphia
- Nephrodesmus
- Nesphostyles
- Nissolia
- Nogra
- Oberholzeria
- Olneya
- Onobrychis
- Ononis
- Ophrestia
- Orbexilum
- Oreophysa
- Ormocarpopsis
- Ormocarpum
- Ormosia
- Orphanodendron

- Ornithopus
- Oryxis
- Ostryocarpus
- Otholobium
- Otoptera
- Otteya
- Oxylobium
- Oxyrhynchus
- Oxytropis
- Pachyrhizus
- Panurea
- Paracalyx
- Paragoodia
- Paramachaerium
- Parochetus
- Parryella
- Pearsonia
- Pedromelum
- Periandra
- Periopsis
- Petaladenium
- Peteria
- Petteria
- Phaseolus
- Phylacium
- Phyllodium
- Phyllota
- Phylloxylon
- Physostigma
- Pickeringia
- Pictetia
- Piptanthus
- Piscidia
- Pisum
- Plagiocarpus
- Platycelyphium
- Platycyamus
- Platylobium
- Platymiscium
- Platypodium
- Platysepalum
- Podolyria
- Podocytisus
- Podolobium
- Poecilanthe
- Poiretia
- Poitea
- Polhillia
- Pongamiopsis
- Pseudarthria
- Pseudeminia
- Pseudoeriosema
- Pseudovigna
- Psophocarpus
- Psoralea
- Psoralidium
- Psorothamnus
- Pterocarpus
- Pterodon
- Ptycholobium
- Ptychosema
- Pueraria
- Pultenaea
- Pycnospora
- Pyranthus
- Rafnia
- Ramirezella
- Ramorinoa
- Retama
- Rhodopsis
- Rhynchosia
- Rhynchotropis
- Riedelella
- Robinia
- Robynsiophyton
- Rothia
- Rupertia
- Sakoanala

- Salweenia
- Sarcodum
- Sartoria
- Schefflerodendron
- Scorpiurus
- Sellocharis
- Sesbania
- Shuteria
- Sigmoidotropis
- Sinodolichos
- Smirnowia
- Smithia
- Soemmeringia
- Sophora
- Spartium
- Spartocytisus
- Spathionema
- Spatholobus
- Sphaerolobium
- Sphaerophysa
- Sphenostylis
- Sphinctospermum
- Spirotropis
- Spongiocarpella
- Stauracanthus
- Staminodianthus
- Steinbachiella
- Streptonanthus
- Stonesiella
- Streblorrhiza
- Strongylodon
- Strophostyles
- Stylosanthes
- Styphnolobium
- Swainsona
- Swartzia
- Sweetia
- Sylvichadsia
- Sysmatium
- Tabaroa
- Tadehagi
- Taralea
- Taverniera
- Templetonia
- Tephrosia
- Teramnus
- Teyleria
- Thermopsis
- Thinicola
- Tipuana
- Trifedacanthus
- Trifolium
- Trigonella
- Tripodion
- Trischedium
- Uleanthus
- Ulex
- Uraria
- Uribea
- Urodon
- Vandasina
- Vatairea
- Vataireopsis
- Vatovaea
- Vavilovia
- Vermifrux
- Vicia
- Vigna
- Vinsenaria
- Vergilia
- Wajira
- Weberbauerella
- Wiborgia
- Wiborgiella
- Wisteria
- Xanthocercis

→ Zollernia
→ Zornia
→ Zygocarpum.

Rosaceae

Br ⊕ ⚥ K(5), C 5, A∞, G ∞ or 1 or (2-5)

→ Great variation in the habit of plants. May be trees, shrubs or trees. Very often they or horny or sometimes climbing.

→ Bear branched tap root

→ Stem is erect or creeping, herbaceous or woody, cylindrical and branched; Many are spiny.

→ Leaves — Alternate, simple or pinnately compound, petiolate, usually stipulate, the stipules are adnate to the petiole, leaf base - conspicuous, small spines often noted on rachis; Margin — Serrate or entire

→ Great variety of inflorescence is noted among species — Corymb, corymbose, umbellate, racemose (or) the flowers may be solitary or in small groups of 2 or 3. Sometimes inflorescence may be Panicle (compound raceme)

→ Flowers hermaphrodite, rarely unisexual (Spiraea aruncus), actinomorphic, regular, bracteolate

→ It may be hypogynous (*Fragaria*), Perigynous (*Prunus*) or epigynous (*Pyrus*)

→ Calyx consists of 5 sepals which are connate at the base (gamosepalous), basal portions are usually adnate forming a hypanthium; calyx lobes are free, valvate or imbricate and green; Bracteoles sometimes form epicalyx; epicalyx lobes remain alternate to the sepals.

→ Corolla consists of 5 petals (rarely four petals) and are polypetalous; Petals usually arise from the rim of hypanthium; imbricate, hypanthium bears a nectariferous glandular disc. Number of petals increases sufficiently in cultivated varieties because of the conversion of stamens into petals. They are variously coloured.

→ Androecium bears indefinite stamens (15-60), sometimes they are 5-10. The stamens generally arranged in one to many whorls of 5 each; they are perigynous around the gynoecium and arise from hypanthium, free; anthers are small, dithecous (two celled), introrse and dorsifixed; dehiscence occurs by means of longitudinal splits; filaments are usually incurved in buds.

→ The number of carpels in Gynoecium is one to many; gynoecium consists of either one compound carpel (syncarpous) or many simple carpels (apocarpous) arranged in cyclic or spiral way. Compound ovary; the ovary is either superior or inferior or half superior half inferior

(Perigynous), 2-5 locules are found when syncarpous; axile placentation and the stigmatic lobes are as many as the number of carpels. The placentation is basal when one carpel is present (apocarpous); ovules are one to many in each carpel. The style is free or connate; stigma - simple, lobed or capitate.

→ Fruit may be dry or fleshy. May be pome, pyriform berry, an etaerio of achenes or one seeded drupe.

→ Seeds are exalbuminous generally with a small embryo.

Classification (Bentham and Hooker)

    Phanerogams
    Dicotyledons
    Polypetalae
    Calyciflorae
    Rosales
    Rosaceae

List of Genus in the family Rosaceae

- Acaena
- Adenostoma
- Agrimonia
- Alchemilla
- Amelanchier
- Aphanes
- Aremonia
- Aria
- Aruncus
- Bencomia
- Brachycaulos
- Cerocarpus

- Chaenomeles
- Chamaebatia
- Chamaebatiaria
- Chamamyeles
- Chamaemespilus
- Chamaerhodes
- Cleffortia
- Coleogyne
- Coluria
- Gillenia
- Hagenia
- Hesperomeles
- Heteromeles
- Holodiscus
- Horkelia
- Horkeliella
- Ivesia
- Kageneckia
- Kelseya
- Kerria
- Leucosidea
- Lindleya
- Luetkea
- Lyonothamnus
- Maddenia
- Malacomeles
- Malus
- Margyricarpus
- Mespilus
- Neillia
- Neviusia
- Nuttalia
- Oemleria
- Orthurus
- Osteomeles
- Pentactina
- Peraphyllum
- Petrophytum
- Photinia
- Physocarpus
- Polylepsis
- Potanina
- Potentilla
- Poterium
- Prinsepia
- Prunus
- Pseudocydonia
- Purshia
- Pyracantha
- Pyrus
- Rhaphiolepis
- Rhodotypos
- Rosa
- Rubus
- Sanguisorba
- Sarcopoterium
- Sibbaldia
- Sibiraea
- Sorbus
- Spenceria
- Spiraea
- Spiraeanthus
- Stephanandra
- Taihangia
- Tetraglochin
- Torminalis
- Vauquelinia
- Waldsteinia
- Xerospiraea

Chapter 62: Saxifrageae

Classification (Bentham and Hooker)

- Phanerogams
- Dicotyledons
- Polypetalae
- Calyciflorae
- Rosales
- Saxifrageae

General characters

→ Plants succulent or non succulent; Annual or Perennial with a basal aggregation of leaves.

→ Leaves — alternate or rarely opposite, usually spiral, herbaceous or fleshy; Petiolate to sessile; Simple or compound

→ Plants hermaphrodite; Floral nectaries usually present; Nectar secretion from the disk; Pollination entomophilous.

→ Inflorescence — cymes, racemes, spikes, heads, fascicles or panicles; rarely solitary.

→ Flowers — Pentamerous, cyclic

→ Perianth with distinct calyx and corolla or rarely sepaline; usually 10 in 2 whorls; isomerous

→ Calyx consists of 5 sepals in 1 whorl; Polysepalous or gamosepalous; regular; Aestivation — imbricate

valvate.
→ Corolla consists of 5 petals in 1 whorl; Poly or gamopetalous; aestivation – imbricate or valvate; regular; May be white, yellow, red or pink. Petals clawed or rarely sessile.
→ Androecium consists of usually 10 stamens, rarely 5, free of the perianth and free of one another; 2 whorled; Stamens – diplostemonous or in some cases isomerous with the perianth. Anthers basifixed or dorsifixed; versatile; dehiscing via longitudinal slits.
→ Gynoecium consists of 2-(5) carpels. apocarpous to syncarpous; Superior to inferior; ovary 2-3 locular; Styles 2-3; free, apical; Stigmas – dorsal to the carpels; Placentation axile; ovules 9-30 per locule arranged in several rows.
→ Fruit – non fleshy; rarely aggregate; dehiscent capsule. Fruit 20-50 seeded
→ Seeds endospermic

Genera included under Saxifrageae

→ Astilbe
→ Astilboides
→ Bensoniella
→ Bergenia
→ Bolandra
→ Boykinia
→ Chrysosplenium
→ Conimitella

- Darmera
- Elmera
- Heuchera
- Jepsonia
- Leptarrhena
- Lithophragma
- Mitella
- Mukdenia
- Oresitrophe
- Quintinia
- Rodgersia
- Saxifraga
- Saxifragella
- Saxifragodes
- Saxifragopsis
- Suksdorfia
- Sullivantia
- Tanakaea
- Tellima
- Tiarella
- Tolmiea

Chapter: 63 Crassulaceae

Classification (Bentham and Hooker)

- Phanerogams
- Dicotyledons
- Polypetalae
- Calyciflorae
- Rosales
- Crassulaceae

General characters

→ Succulent herbs or shrubs; Perennial

→ Leaves - alternate or opposite or whorled; when alternate, spiral; flat or terete; fleshy; petiolate to sub sessile; Lamina usually entire or rarely sessile; exstipulate

→ Plants hermaphrodite; floral nectaries present; Pollination entomophilous

→ Inflorescence - cymes or corymbs; Solitary.

→ Flowers - regular, 3-5 merous, cyclic. Floral receptacle not markedly hollowed. Free hypanthium present.

→ Perianth with distinct calyx and corolla

- Calyx – 3 to 30 in 1 whorled; Polysepalous or rarely gamosepalous; regular, Persistent, imbricate.
- Corolla – 3 to 30 petals in ..1 whorl; usually Polypetalous rarely polypetalous.; Corolla lobes Markedly longer than tube; Corolla – imbricate; regular, white/yellow/Pink/Purple
- Androecium consists of 3–30 Stamens, free of the perianth or rarely adnate; usually free of one another or rarely basally coherent – Monoadelphous; usually 2 whorled (rarely 1) Stamens – diplostemonous; Anthers more or less basifixed; dehiscing via longitudinal slits.
- Gynoecium consists of 3–30 carpels isomerous with the Perianth; apocarpous; superior; apically stigmatec; style short or long; 1–50 Ovuled; Placentation – sub Marginal
- Fruit – non fleshy, aggregate
- Seeds – oily endosperm; straight embryo.

Genera included under Crassulaceae

- Adromisschus
- Aeonium
- Aichryson
- Bryophyllum
- Chiastophyllum
- Cotyledon

- Crassula
- Cremnophila
- Diamorpha
- Dudleia
- Echeveria
- Graptopetalum
- Greenovia
- Hylotelephium
- Hypagophytum
- Jovibarba
- Kalanchoe
- Lenophyllum
- Meterostachys
- Monanthes
- Mucizonia
- Orostachys
- Pachyphytum
- Pagella
- Parvisedum
- Perrierosedum
- Pistorinia
- Pseudosedum
- Rhodiola
- Rochea
- Rosularia
- Sedum
- Sempervivum
- Sinocrassula
- Telmissa
- Thompsonella
- Tylecodon
- Umbilicus
- Villadia

Chapter – 64
Droseraceae

Classification (Benthany and Hooker)

- Phanerogams
- Dicotyledons
- Polypetalae
- Calyciflorae
- Rosales
- Droseraceae

General Characters

→ Mostly perennial creeping insectivorous herbs of marshy places with underground stems. Primary root is absent in some species, the place of which is taken by protocorm-like structure developed from hypocotyl and bears filiform attaching organs.

Based on climatic conditions, two types of growth forms exist — Photophilous form, in which the stem is capable of indefinite growth by means of terminal bud, and geophilous form are exhibited by species living in a condition where damp cold season alternates with dry hot season and they persist by means of swollen root or bulb.

Generally primary root does not develop, instead

numerous adventitious roots are formed.

→ Leaves :- alternate, very rarely whorled, lamina covered with tentacular glands, stipules mostly present, membranous. Leaves are modified for catching insect preys and for photosynthesis. Petiolate. <u>Dionaea</u> - Petiole is winged and acts like a leaf. Upper surface of leaf is provided with tentacles which are glandular hairs.

Morphologically a tentacle is intermediate between a true leaf part and a trichome. A tentacle is provided with vascular chord, a tracheid end remains covered by two layers of secreting layer made up of secreting epithelium. The tentacles play an important role for catching tiny insects. The tentacles have peculiar power of movement. As soon as an insect gets stuck into the mucilagenous secretions of glands, the tentacles bend down over the body of the prey and catch them. The insects die, and the proteins are absorbed by the tentacles.

→ Inflorescence - may be solitary or in spike or raceme like inflorescence, or scorpiod cyme (cincinnus), racemose to paniculate.

→ Flowers - bisexual, actinomorphic, hypogynous, tetramerous or pentamerous, perianth biseriate, distinguishing into sepals and petals.

→ Calyx consists of 4-5 sepals, mostly briefly and basally connate, persistent, imbricate.

→ Corolla consists of 5 petals, distinct, convolute, imbricate, membranous or soft.

→ Androecium consists of 5-20 stamens, in one or more pentamerous whorls, distinct or rarely the filaments basally connate, anthers dithecous, the cell sometimes divergent and with broad connective, dehisce longitudinally, extrorse.

→ Gynoecium - 3-5 carpels, syncarpous, superior, 1 locule, styles 3-5, mostly free, stigma capitate; ovules many, sometimes few, anatropous, placentation parietal, rarely on free central placenta.

→ Fruit - A loculicidal capsule.

→ Seed - numerous, small, embryo straight, small cotyledons.

Genera included under Droseraceae

→ Dionaea
→ Aldrovanda
→ Drosera

Chapter-65: Hamamelideae

Classification (Bentham and Hooker)

Phanerogams
~~Dicotyledons~~ Dicotyledons
Polypetalae
Calyciflorae
Rosales
Hamamelideae

General characters

→ Trees and shrubs; often with stellate indumentum
→ Leaves - persistent or deciduous; alternate; spiral or distichous; petiolate; simple; lamina - dissected or entire; stipulate.
→ Plants hermaphrodite or monoecious; Pollination anemophilous or entomophilous.
→ Inflorescence - spikes (usually), heads or racemes or panicles. Involucral bracts may be present in some cases.
→ Perianth with distinct calyx and corolla. but often sepaline
→ Calyx consists of 4-5 sepals in 1 whorl; poly or gamosepalous; regular; imbricate
→ Corolla when present consists of 2-4 or 5 petals

in 1 whorl; gamopetalous; valvate or with open aestivation; Petals clawed or sessile
→ Androecium consists of 4-5 or 10-14 or 15-32 stamens; free of the perianth; free from one another; 1 or 2 whorled; usually basifixed (anthers); introrse
→ Gynoecium - bi or tri carpellary; apocarpous to syncarpous; superior to inferior; ovary 2 locular; Style - 2, often recurved; free to partially joined; apical; Stigma - dry type. Placentation axile; ovules 1-15 per locule
→ Fruit - non fleshy, dehiscent, capsule
→ Seeds endospermic (or!); winged or wingless; Embryo straight

Genera included under Hamamelideae

- → Chunia
- → Corylopsis
- → Dicoryphe
- → Disanthus
- → Distyliopsis
- → Distylium
- → Embolanthera
- → Eustigma
- → Exbucklandia
- → Fortuneania
- → Fothergilla
- → Hamamelis
- → Loropetalum
- → Maingaya
- → Matudaea
- → Molinadendron
- → Mytilaria
- → Neostrearia
- → Noahdendron
- → Ostrearia
- → Parrotia
- → Parrotiopsis
- → Sinowilsonia
- → Sycopsis
- → Tetrathyrium
- → Trichocladus

Chapter-66: Bruniaceae

Classification (Bentham and Hooker)

    Phanerogams
    Dicotyledons
    Polypetalae
    Calyciflorae
    Rosales
    Bruniaceae

General characters

→ Ericoid shrubs or rarely undershrubs or trees.

→ Leaves – small; alternate, spiral; imbricate; non-sheathing, simple; Lamina – entire; Parallel veined; Leaves exstipulate

→ Plants hermaphrodite

→ Flowers rarely solitary; usually aggregated in inflorescence – spikes or heads.

→ Flowers – 5 bracteate, small or rarely medium sized, regular; Tetra or pentamerous; tetracyclic. Free hypanthium may be present or absent; Hypogynous disc when present – intrastaminal.

→ Perianth with distinct calyx and corolla

→ Calyx consists of 4-5 sepals in 1 whorl;

Families of Dicotyledons

- Poly or gamosepalous; regular, persistent, imbricate
- Corolla consists of 4-5 petals; poly or gamopetalous; imbricate, regular; often persistent, in some cases deciduous; petals often clawed or rarely sessile.
- Androecium – 4 to 5 stamens; free of the perianth or adnate to the claws of the petals of the petals to form a tube; 1 whorl; Stamens – isomerous with perianth, opposite sepalous; anthers – sessile or filantherous; anthers dorsifixed; often versatile; dehiscing via longitudinal slits; introrse
- Gynoecium – usually bicarpellary, rarely tricarpellary; Gynoecium – monomerous or syncarpous; partly inferior to inferior or rarely superior; Styles – 2-3, usually partially joined; apical. Stigma – 2-3. Placentation apical; Ovules 2-12 per locule
- Fruit – indehiscent, an achene, or nut
- Seeds – endospermic; straight embryo.

Genera included under Bruniaceae

- Audouinia
- Berzelia
- Brunia
- Linconia
- Lonchostoma
- Mniothamnea
- Nebelia
- Pseudobaeckea
- Raspalia
- Staavia
- Thamnea
- Tittmannia

Chapter 67
Haloragaceae

Classification (Bentham and Hooker)

- Phanerogams
- Dicotyledons
- Polypetalae
- Calyciflorae
- Rosales
- Halorageae

General characters

→ Mostly herbs or shrubs; hydrophytic to halophytic or mesophytic.
→ Leaves of hydrophytes submerged or emergent
→ Heterophylly – submerged leaves are dissected and the emergent leaves are more or less entire.
→ Leaves – alternate or opposite or whorled. When alternate – spiral; Petiolate to sessile; Simple or compound; Lamina dissected or entire; Exstipulate
→ Plants hermaphrodite or monoecious Pollination – anemophilous.

→ Flowers – Solitary; inflorescence – Spikes, corymbs or racemes.
→ Flowers – bracteolate or rarely ebracteolate; minute; regular; 2-4 merous, tetra or pentacyclic
→ Perianth may be with distinct calyx or corolla; or sequentially intergrading from sepals to petals or vestigial to absent.
→ Calyx consists of 2 or 4 sepals in 1 whorl; Polysepalous; regular; Persistent; valvate
→ Corolla consists of 2 or 4 petals in 1 whorl; Polypetalous; regular
→ Androecium consists of 8 or rarely 3-4 stamens; free of the perianth and free of one another; 1 or 2 whorled; Stamens – Isomerous with perianth or diplostemonous; Anthers basifixed; dehisce via longitudinal slits
→ Gynoecium – 2-3 or 4 carpelled; syncarpous; inferior ovary – 1-4 locular; Styles 2-4, feathery, free, apical. Stigma – dry type. Placentation apical; Ovules 1 per locule; anatropous
→ Fruit – non fleshy, indehiscent or a schizocarp (drupe or nut)
→ Seeds – copiously endospermic; Endosperm oily;

Embryo straight

<u>Genera included under Haloragaceae</u>

→ Glischrocaryon
→ Gonocarpus
→ Haloragis
→ Haloragodendron
→ Laurembergia
→ Meziella
→ Myriophyllum
→ Proserpinaca.

Sub class: polypetalae

Series: calyciflorae

Order: Myrtales

Families:

Rhizophoraceae

Combretaceae

Myrtaceae

Melastomaceae

Lythrarieae

Onagrarieae

Chapter - 68
Rhizophoraceae

Classification (Bentham and Hooker)

- Phanerogams
- Dicotyledons
- Polypetalae
- Calyciflorae
- Myrtales
- Rhizophoraceae

General characters

→ Trees or shrubs
→ Leaves opposite or whorled; leathery; Petiolate; simple; lamina entire; stipulate; stipules interpetiolar; Lamina margin - entire, crenate or dentate.
→ Plants hermaphrodite but rarely unisexual flowers present.
→ Inflorescence - cymes, racemes or fascicles (or) rarely solitary, axillary.
→ Flowers - regular, 3-4-6-20 merous, free hypanthium present or absent. Hypogynous disk when present, intrastaminal.

→ Perianth with distinct calyx and corolla.
→ Calyx consists of 3-16 sepals, 1 whorled, polysepalous, regular, fleshy, persistent, valvate
→ Corolla consists of 3-16 petals, 1 whorled, polypetalous, contorted; fleshy; petals clawed or sessile;
→ Androecium consists of 8-40 stamens; free of the perianth; free of one another or coherent (basally connate); 1 whorled; sometimes stamens bundled; diplostemonous to polystemonous; filantherous or sessile anthers; Anthers introrse.
→ Gynoecium – 2-20 carpelled; pistil – 1 to 20 celled; syncarpous; superior to inferior ovary; Ovary 1 locular or 2-20 locular (locules with false septa); Styles-1, apical; Stigma-1, lobed or capitate, generally papillate; Placentation apical or axile; Ovules 2 per locule or 3-25 per locule.
→ Fruit – usually fleshy; rarely non-fleshy; dehiscent or indehiscent; capsule or berry or

drupe.

→ Seeds with copious endosperm (oily); Seeds - winged or wingless, Embryo straight

Genera included under Rhizophoraceae

- → Anopyxis
- → Blepharistemma
- → Bruguiera
- → Carallia
- → Cassipourea
- → Ceriops
- → Comiphyton
- → Crossostylis
- → Dactylopetalum
- → Gynotroches
- → Kandelia
- → Macarisia
- → Pellacalyx
- → Rhizophora
- → Sterigmapetalum
- → Weihea.

Chapter - 69
Combretaceae

Classification (Bentham and Hooker)

Phanerogams
Dicotyledons
Polypetalae
Disciflorae
Myrtales
Combretaceae

General Characters:

→ Trees, shrubs and climbers
→ Leaves - simple, alternate, sub-opposite or opposite, rarely ternate, petiolate, entire, exstipulate
→ Inflorescence - Spikes or Racemes, racemes often panicled
→ Flower - bracteolate, hermaphrodite or unisexual by reduction, actinomorphic with a tendency to zygomorphy, pentamerous, rarely tetramerous, epigynous.
→ Calyx consists of 4-5 lobes, the tube remains adnate to the ovary and is produced above it. Valvate aestivation.
→ Corolla consists of 4-5 petals alternating the sepals or altogether absent.

→ Androecium – the stamen numbers from 4–5 or 8–9, situated on the calyx; generally the stamen number is twice the number of sepals; found in 2 series, the lower opposite the sepals and the upper alternating the sepals; in bud condition, stamens are bent inward.

→ Gynoecium – monocarpellary, ovary one chambered, inferior, angled, anatropous ovules (2–5 or more); style simple, one, long, filiform bearing a simple pointed or rarely capitate stigma.

→ Fruit – 2 to 5 angled or winged, coriaceous or drupaceous, one seeded, indehiscent.

–) Seeds are non endospermic (ex.alburninous), the cotyledons are spirally folded.

→ Pollination to entomophilous.

Genera included under Combretaceae

- Anogeissus
- Buchenavia
- Bucida
- Calopyxis
- Calycopteris
- Combretum
- Conocarpus
- Dansiea
- Guiera
- Laguncularia
- Lumnitzera
- Macropteranthes
- Melostemon
- Pteleopsis
- Quisqualis
- Strephonema
- Terminalia
- Terminaliopsis
- Thiloa

Chapter - 70
Myrtaceae

Classification (Bentham and Hooker)

Phanerogams
Dicotyledons
Polypetalae
Discliflorae
Myrtales
Myrtaceae

General Characters

→ Plants are moderate sized or small trees or shrubs, herbs are rare.
→ Branched tap root system
→ Stem - Erect, branched, woody, solid.
→ Leaves - Mostly opposite, sometimes alternate (Eucalyptus). Petioles are short. Leaves - simple, entire, evergreen, exstipulate. Many xerophytic plants possess acicular (needle-like) leaves. Leaves contain numerous oil glands which secrete ethereal oils.
→ Inflorescence - cymose type. Some flowers are solitary in leaf axils (Myrtus communis). Rarely inflorescence is of racemose type. Panicled cyme in Syzgium fruticosum. Corymbose type in S. mappaceum. In Psidium guajava 1-3 flowers are found in each axil. Callistemon, flowers are arranged in spikes. Flowers are arranged in axillary umbels in Eucalyptus teretocornis

→ Flowers – bracteate, bracteolate, actinomorphic, hermaphrodite, regular, complete, cyclic, epi or perigynous. Receptacle is either united forming a complete epigynous condition (*Myrtus communis*) or the union may not be completed resulting in perigynous condition.

→ Calyx consists of either 4 or 5 sepals, alternating with 4 or 5 free petals. Calyx is thrown off like a cap when the flower opens (*Eucalyptus*). In some, sepals are more or less united. Calyx is much reduced in some cases. Aestivation is imbricate.

→ Corolla consists of 4 or 5 petals, free (polypetalous) and more or less circular in form. Petals are united to form a cap in *Eucalyptus*. Aestivation is imbricate or quincuncial.

→ Androecium consists of indefinite number of stamens which are free and arranged in whorls on the edge of the receptacle. Rarely they are arranged in obdiplostemonous whorls. Filaments bent inwards in the bud and may be free or more or less coherent at the base. Anthers are dorsifixed or versatile, introrse and dehisce via longitudinal slits. In *callistemon*, numerous long filaments form a scarlet brush many times longer than corolla.

→ Gynoecium consists of 2-α carpels, syncarpous, ovary inferior or half inferior. Ovary contains one to many loculi, 2-α somewhat obliquely pendulous anatropous or campylotropous ovules are found in each locule. Placentation is axile, rarely parietal (_Rhodamnia_). Style is simple, long and generally flexuose with a capitate or simple stigma.

→ Fruit - May be a berry or very rarely a drupe. Sometimes loculicidal capsule (_Callistemon_)

→ Seed - Exalbuminous, with straight or more or less spirally rolled or bent embryo.

→ Pollination is entomophilous.

Genera included under Myrtaceae

- Acca
- Accara
- Acmena
- Acmenosperma
- Actinodium
- Agonis
- Allosyncarpia
- Amomyrtella
- Amomyrtus
- Angasomyrtus
- Angophora
- Aphanomyrtus
- Archirhodomyrtus
- Arillastrum
- Astartea
- Asteromyrtus
- Austromyrtus
- Backhousia
- Baeckea
- Balaustion
- Borangia
- Basisperma
- Beaufortia
- Blepharocalyx
- Callistemon
- Calothamnus
- Corynanthera
- Corynemyrtus
- Cupheanthus
- Darwinia
- Decaspermum
- Eremaea
- Eucalyptopsis
- Eucalyptus
- Eugenia
- Gomidesia
- Feijoa
- Hexachlamys
- Homalocalyx
- Homalospermum
- Homoranthus
- Hottea
- Hypocalymma
- Jambosa
- Kania
- Kjellbergiodendron
- Kunzea
- Lamarchea
- Legrandia
- Lencymmoea
- Leptospermum
- Lindsayomyrtus
- Lophomyrtus
- Lophostemon
- Luma
- Lysicarpus
- Malleostemon
- Marlieria
- Melaleuca
- Meteoromyrtus
- Metrosideros
- Micromyrtus
- Mitranthes
- Mitrantia
- Monimiastrum
- Mosiera
- Mozartia
- Myrceugenia
- Myrcia
- Myrcianthes
- Myrciaria
- Myrrhinium
- Myrtastrum
- Myrtella
- Myrteola

- Myrtus
- Neofabricia
- Neomitranthes
- Neomyrtus
- Ochrosperma
- Octamyrtus
- Osbornia
- Paramyrciaria
- Pericalymma
- Phymatocarpus
- Pileanthus
- Piliostigma
- Piliocalyx
- Pimenta
- Pleurocalyptus
- Plinia
- Pseudanamomis
- Pseudeugenia
- Psidium
- Purpureostemon
- Pyrenocarpa
- Regelia
- Rhodamnia
- Rhodomyrtus
- Rinzia
- Ristantia
- Rylstonea
- Scholtzia
- Siphoneugenia
- Sphaerantia
- Stereocaryum
- Syncarpia
- Syzygium
- Tepualia
- Thryptomene
- Tristania
- Tristaniopsis
- Ugni
- Uromyrtus
- Verticordia
- Waterhousea
- Welchiodendron
- Whiteodendron
- Xanthomyrtus
- Xanthostemon.

Chapter-71
Melastomataceae

Classification (Bentham and Hooker)

- Phanerogams
- Dicotyledons
- Polypetalae
- Calyciflorae
- Myrtales
- Melastomataceae

General characters

→ Herbs or shrubs or trees or lianas

→ Leaves opposite or rarely whorled; simple; lanceolate or ovate; lamina entire; exstipulate. Margin- entire or serrate

→ Plants hermaphrodite; floral nectaries may or may not be present; pollination- entomophilous, ornithophilous or cheiropterophilous

→ Flowers rarely solitary or usually aggregated in inflorescence cymose

→ Flowers - bracteolate (bracteoles often brightly coloured); calyptrate; regular or irregular (irregularity involving the androecium). Flowers - 3-5-7 merous; cyclic; free hypanthium present (tubular or campanulate)

→ Perianth with distinct calyx and corolla;

→ Calyx consists of 4-7 sepals, 1 whorled, gamosepalous (sometimes united forming a calyptra), entire or lobulate or blunt lobed; regular; imbricate, valvate or contorted.

→ Corolla - 4-7 petals in 1 whorled; usually polypetalous; contorted; regular

→ Androecium - 4 to 96 stamens, free of one another and free of the perianth; 1 whorled or 2 whorled; Anthers - basifixed; dehisce via longitudinal slits.

→ Gynoecium consists of 3-14 carpels, syncarpous; superior to inferior; ovary 3-14 locular; Styles 1 apical; Stigma - 1; placentation usually axile but median parietal when unilocular.

→ Fruit - fleshy or non fleshy, dehiscent or indehiscent; capsule or berry.

→ Seeds - non-endospermic

Genera included under Melastomaceae

- → Acanthella
- → Aciotis
- → Acisanthera
- → Adelobotrys
- → Allomaceta
- → Allomorpheia
- → Allomorphia
- → Amphiblemma
- → Amphorocalyx
- → Anaectocalyx
- → Anerincleistus
- → Antherotoma
- → Appendicularia
- → Anthrostemma
- → Ascistanthera
- → Astrocalyx
- → Astronia
- → Astronidium

Families of Dicotyledons

- → Axinaea
- → Barthea
- → Beccarianthus
- → Behuria
- → Bellucia
- → Benevidesia
- → Bertolonia
- → Bisglaziovia
- → Blakea
- → Blastus
- → Boerlagea
- → Boyania
- → Brachyotum
- → Bredia
- → Brittenia
- → Bucquetia
- → Cailliella
- → Calvoa
- → Calycogonium
- → Cambessedesia
- → Campimia
- → Carionia
- → Castratella
- → Catanthera
- → Catocoryne
- → Centradenia
- → Centradeniastrum
- → Centronia
- → Chaetolepis
- → Chaetostoma
- → Chalybea
- → Charianthus
- → Cincinnobotrys
- → Clidemia
- → Comolia
- → Comoliopsis
- → Conostegia
- → Creochiton
- → Cyanandrium
- → Cyphostyla
- → Cyphotheca
- → Dalenia
- → Desmoscelis
- → Dicellandra
- → Dichaetanthera
- → Dinophora
- → Dionycha
- → Dionychastrum
- → Diplarpea
- → Diplectria
- → Dissochaeta
- → Dissotis
- → Dolichoura
- → Driessenia
- → Enaulophyton
- → Eriocnema
- → Ernestia
- → Feliciadamia
- → Fordiophyton
- → Fritzschia
- → Graffenriedia
- → Gravesia
- → Guyonia
- → Henriettea
- → Henriettella
- → Heterocentron
- → Heterotis
- → Heterotrichum
- → Huberia
- → Huilaea
- → Hypenanthe
- → Kendrickia
- → Kerriothyrsus
- → Killipia
- → Kirkbridea

Families of Dicotyledons

- → Lavoisiera
- → Leandra
- → Lithobium
- → Llewelynia
- → Loreya
- → Loricalepis
- → Macairea
- → Macrocentrum
- → Maciolenes
- → Maguireanthus
- → Maieta
- → Mallophyton
- → Marcetia
- → Mecranium
- → Medinilla
- → Melastoma
- → Melastomastrum
- → Meriania
- → Merianthera
- → Miconia
- → Microlepis
- → Microlicia
- → Monimsenia
- → Monochaetum
- → Monolena
- → Myriaspora
- → Myrmidone
- → Neblinanthera
- → Necranium
- → Neodriessenia
- → Nepsera
- → Nerophila
- → Ochthephilus
- → Ochthocharis
- → Omphalopus
- → Opisthocentra
- → Oritrephes
- → Osbeckia
- → Ossaea
- → Otanthera
- → Oxyspora
- → Pachyanthus
- → Pachycentria
- → Pachyloma
- → Phainantha
- → Phyllagathis
- → Pilocosta
- → Plagiopetalum
- → Pleiochiton
- → Plethiandra
- → Pogonanthera
- → Poikilogyne
- → Poilannammia
- → Pomatostoma
- → Poteranthera
- → Preussiella
- → Pseudodissochaeta
- → Pseudosbeckia
- → Pterogastra
- → Pterolepis
- → Pyramia
- → Rhexia
- → Rhynchanthera
- → Rousseauxia
- → Salpinga
- → Sandemania
- → Sarcopyramis
- → Schwackaea
- → Scorpiothyrsus
- → Siphanthera
- → Sonerila
- → Sporoxeia
- → Stenodon
- → Stussenia
- → Svitramia
- → Tateanthus

- Tayloriophyton
- Tessmannianthus
- Tetrazygia
- Tibouchina
- Tibouchinopsis
- Tigridiopalma
- Tococa
- Topobea
- Trembleya
- Triolena
- Tristemma
- Tryssophyton
- Tylanthera
- Vietsenia

Chapter-72
Lythraceae

Classification (Bentham and Hooker)

- Phanerogams
- Dicotyledons
- Polypetalae
- Disciflorae
- Myrtales
- Lythraceae

General characters

→ herbs, shrubs or trees.

→ Leaves – Opposite or sometimes whorled, exstipulate or minutely stipulate, entire and simple; net veined, petiolate, apex acute

→ Inflorescence – cymose panicles or solitary

→ Flower – Actinomorphic, bisexual, regular and perigynous; flowers subtended by epicalyx like united pair of bracteoles (sepal stipule). Rarely zygomorphic flowers occur (cuphea).

→ Perianth is usually 4-6 merous however in Lafoensia there are 8-16 merous perianth. Petals absent in Peplis and some species of Ammannia

→ Androecium – stamens usually twice the number of petals in 2 whorls or many; those of the outer whorl alternate with petals and are inserted on the inside of the receptacle cup or hypanthium much lower down, filaments unequal; sometimes stamens of the inner whorl are absent; anthers dithecous, dehiscence longitudinal, introrse. <u>Rotala</u> - 1 stamen only.

→ Gynoecium - 2-6 carpels, united in a superior or semi-inferior 2-6 celled ovary with simple style and capitate stigma; ovules anatropous, usually many on axile placentation; sometimes partition septa disappear in the inner part of the ovary.

→ Fruit - A capsule; seed with straight embryo; non-endospermic.

Characteristic features of the family ⇒ This family is distinguished by the presence of hypanthium, superior ovary, crumpled corolla, the often unequal stamens of typically twice the number of sepals, and the seed without endosperm. Intraxylary phloem, mucilagenous cells (in leaf) and secretory cavities are present.

Genera included under Lythraceae

- Adenaria
- Ammannia
- Capuronia
- Crenea
- Cuphea
- Decodon
- Didiplis
- Diplusodon
- Galpinia
- Ginoria
- Haitia
- Heimia
- Hionanthera
- Koehneria
- Lafoensia
- Lagerstroemia
- Lawsonia
- Lourtella
- Lythrum
- Nesaea
- Pehria
- Pemphis
- Peplis
- Physocalymma
- Pleurophora
- Rotala
- Tetrataxis
- Woodfordia

Chapter - 73
Onagraceae

Classification (Bentham and Hooker)

Phanerogams
Dicotyledons
Polypetalae
Discflorae
Myrtales
Onagraceae.

General Characters

→ Mostly herbs, occasionally aquatic, rarely shrubs (*Fuchsia*) or trees (*Hauya*)

→ Leaves – alternate or opposite, simple and exstipulate. In some genera (*Epilobium*), a large leaf similar in appearance like adult leaf develops as a result of intercalary growth below the cotyledons which are borne above the ground during germination. Deciduous stipules are found in *Fuchsia* and *Circaea*

→ Inflorescence – Spike, racemose or solitary axillary, Paniculate in *Fuchsia*

→ Flowers – bisexual, actinomorphic or infrequently zygomorphic, typically tetramerous, epigynous, complete and regular. The flower of *Trapa* is semi epigynous with half inferior ovary of two carpels and single pendulous ovule in each carpellary chamber.

→ Calyx consists of 4 sepals, valvate, forming the calyx tube being basally united and ultimately adnate to the hypanthium, sometimes persistent (_Jussieua_). Persistent tepals assume spinous structure above the fruit in some cases (_Trapa_).

→ Corolla usually 4 petals, sometimes 2 or more or less none, free and imbricate; mostly clawed and convolute. In _Lopezia_, two posterior petals are smaller than antero-lateral ones and are bent upwards a little from the base.

→ Androecium - Stamen numbers are usually same as that of petals or in 2 whorls and twice as many; 1 fertile stamen and 1 staminode in _Lopezia_; when biseriate, those of the outer whorls alternate with petals, distinct, arising from or near hypanthium rim; anthers typically dithecous, dehiscing longitudinally sometimes each cell divides. Pollen grains in _Epilobium_ are united in tetrads.

→ Gynoecium - carpels 4, rarely 2, 3 or 5-6 united into 4 celled inferior ovary, rarely ovary is semi inferior; style one, simple, slender, stigma usually capitate, sometimes notched; ovules usually several, sometimes few or one, anatropous, ascending, horizontal or descending

on axile placentation.
→ Fruit — loculicidal capsule, berry, drupe or nut
→ Seeds - non-endospermic with small embryo; seeds cymose or glabrous
→ Trapa _ the 2 cotyledons are very unequal in size, one is scaly minute and the other is large with full of food material. The large food containing cotyledon remains within the seed, the other scaly is epigeal which is carried up by hypocotyl and turns green.

Genera included under onagraceae

- Boisduvallia
- Calylophus
- Camissonia
- Circaea
- Clarkia
- Epilobium
- Erenothera
- Fuchsia
- Gaura
- Gayophytum
- Gongylocarpus
- Hauya
- Jussiaea
- Lopezia
- Ludwigia
- Oenothera
- Stenosiphon
- Xylonagra.

Sub class: polypetalae

Series: calyciflorae

Order: Passiflorales

Families:

Samydaceae

Loaseae

Turneraceae

Passifloreae

Cucurbitaceae

Begoniaceae

Datisceae

Chapter-74
Samydaceae

Classification (Bentham and Hooker)

    Phanerogams
    Dicotyledons
    Polypetalae
    Calyciflorae
    Passiflorales
    Samydaceae

General characters

→ Shrubs or trees
→ Leaves - alternate, pinnately veined, rarely acrodromous, entire or teethed; rarely spinose margin; stipulate or exstipulate
→ Hermaphrodite or rarely dioecious or monoecious
→ Inflorescence - axillary fascicles, corymbs, racemes of spikes or panicles of spikes.
→ Sepals 4-7
→ Petals absent
→ Disk usually present, adnate to the calyx and alternating with or inside the whorl of stamens.

- Stamens 4 to numerous in 1-3 whorls; sometimes connate, anthers - introrse or latrorse;
- Gynoecium - 1 pistil, ovary unilocular, superior; Placentation parietal; ovules few to numerous; Styles -1; stigma - usually capitate, rarely sessile
- Fruit - fleshy or dry 3-valved capsules rarely indehiscent
- Seeds - arillate or with long, cottony hairs.

Genera included under Samydaceae
- Tetrathylacium
- Lunania
- Casearia
- Ryania
- Trichostephanus
- Osmelia
- Pseudosmelia
- Ophiobotrys
- Euceraea
- Neoptychocarpus
- Casearia
- Laetia
- Zuelania
- Hecatostemon
- Samyda

Chapter - 75
Loaseae

Classification (Bentham and Hooker)

Phanerogams
Dicotyledons
Polypetalae
Calyciflorae
Passiflorales
Loaseae

General characters

→ Opposite phyllotaxy
→ Reticulate testa
→ Parallel stigmatic lobes
→ Pluriovulate ovaries
→ Polystemonous androecia
→ Petals with numerous veins from base
→ Annular nectaries
→ Flowers - erect with white/yellow/grey petals and pale nectar scales (or) pendent flowers, brightly coloured petals, variously elaborated and coloured nectar scales.
→ Annuals with showy flowers
→ Genera _Loasa_ and _Caiophora_ are covered with

glandular hairs or bristles, which sting much worse.

Genera included under Loaseae

- → Blumenbachia
- → Caiophora
- → Loasa
- → Scyphanthus
- → Chichicaste
- → Huidobria
- → Kissenia
- → Xylopodia
- → Klaprothia
- → Plakothira

Chapter - 76

Turneraceae

Classification (Bentham and Hooker)

Phanerogams
Dicotyledons
Polypetalae
Calyciflorae
Passiflorales
Turneraceae

General Characters

→ Shrubs or herbs; rarely trees
→ Leaves - alternate, spiral, petiolate, simple
→ Lamina - dissected or entire; usually exstipulate often with a pair of glands or extra floral nectaries at the base of the blade rarely stipulate
→ Plants hermaphrodite; Pollination - self pollination or entomophilous
→ Flowers - solitary, axillary or inflorescence - racemes, often bracteate, regular, pentamerous, tetracyclic. Free hypanthium present
→ Perianth with distinct calyx and corolla.
→ Calyx - 5 sepals, 1 whorl, gamosepalous, toothed, regular, imbricate

→ Corolla – 5 petals, 1 whorled; polypetalous; contorted; regular, yellow or red.
→ Androecium – 5, isomerous with perianth, anthers introrse, dehisce via longitudinal slits
→ Gynoecium – tricarpellary, syncarpous, superior to partly inferior; ovary 1 locular; styles – 3, stigmas – 3. Ovules 3–many in single cavity.
→ Fruit – non-fleshy, dehiscent, capsule
→ Seeds – endospermic (oily), embryo straight

Genera included under Turneraceae

- Adenoa
- Erblichia
- Hyalocalyx
- Loewia
- Mathurina
- Piriqueta
- Stapfiella
- Streptopetalum
- Tricliceras
- Turnera

Chapter-77
Caricaceae [Passiflorea]

Classification (Bentham and Hooker)

- Phanerogams
- Dicotyledons
- Polypetalae
- Thalamiflorae
- Passiflorales
- Passiflorea (Caricaceae)

General Characters

→ Small soft wooded tree
→ Stem stout and generally unbranched.
→ Tissues are permeated with a laticiferous system which contains the powerful proteolytic enzyme, Papain.
→ Leaves - large and palmately lobed. Petioles are long and fistular. They are exstipulate and arranged alternately or spirally usually forming a crown at the terminal end of the stem.
→ Flowers - Usually unisexual and rarely bisexual. Usually the flowers are regular and pentamerous. Morphology of the flowers are grouped into 4 types
→ **Male flowers** — These flowers develop on male plants. The flowers are sessile and unisexual.

Sometimes, the terminal flowers of the inflorescence are either female or bisexual. The number of stamens is 10, arranged in 2 whorls of 5 each. Rudimentary pistils may be present. Flowers are arranged in clusters or pendant racemes or richly branched panicles.

→ **Female flowers :-** These flowers develop on female plants and are subsessile. Ovary is large and globose. Style short, stigma fine. Fruit is globose or ovoid. Flowers are arranged in few-flowered corymbs in leaf axils.

→ **Long-fruited flowers :-** The flowers are like female flowers. Corolla is united and 10 stamens present at the base of petals. Ovary is elongated and the fruit is cylindrical in shape. Flowers are arranged in 5-6 flowered corymbs in the axils of the leaf.

→ **Polygamous flowers :-** Such flowers are of two types. One type bear 10 sessile stamens found at the throat of the corolla and the other type bear 5 stamens with long filaments, found at the base of the ovary. Corolla tube is very minute.

→ Calyx consists of 5 sepals, united (gamosepalous). It is very minute in all the flowers.

→ Corolla consists of 5 petals which may be united or free. Aestivation is twisted. Male flowers bear long corolla tube and the female ones the short tube.

→ Androecium consists of 10 stamens arranged in 2 whorls of 5 each. Sometimes the stamens of the inner whorls are suppressed. They are epipetalous. The filaments may be short or long. Anthers are dithecous; dehisce by longitudinal slits.

→ Gynoecium consists of 5 carpels, united - syncarpous. Ovary is unilocular. Placentation is parietal. Ovules are many.

→ Fruit is large ovoid elongated berry.

→ Seeds - Numerous, black coloured seeds which are albuminous. Each seed bears a straight embryo with two large cotyledons.

→ Cross pollination aided by insects.

Genera included under Caricaceae

- Carica
- Cyclicomorpha
- Jacaratia
- Jarilla
- Vasconcellea

Chapter – 78
Cucurbitaceae

Classification (Bentham and Hooker)

    Phanerogams
    Dicotyledons
    Polypetalae
    Thalamiflorae
    Passiflorales
    Cucurbitaceae

General characters:

→ Majority are annual or perennial, trailing or decumbent vines, climbing by means of tendrils. *Acanthrosicyos* is a thorny leafless shrub, two species of *Dendrosicyos* are small trees but the stem are soft.

→ Branched Tap root system

→ Stem – herbaceous, climbing, hollow, often 5 angled and either glabrous, hairy or prickly

→ Bi collateral vascular bundles are present

→ Leaves – simple, mostly palmately lobed or divided, usually exstipulate, large and long petaloid, alternate; petioles often hollow; tendrils may be simple or branched and may twist to the right and to the left at different points along their axes.

→ Some botanists believe tendrils to be leaves or

Some consider it as modification of bracteoles, Some consider as modification of stipules

→ Inflorescence- cymose solitary axillary, sometimes Panicles, more rarely racemose or sub-umbellate

→ Flower:- typically regular and unisexual and either monoecious or dioecious. Actinomorphic, epigynous, Pentamerous but with the tricarpellary Pistil by reduction and the stamens united into two pairs. Calyx and corolla inserted on a variously shaped often elongated hypanthium which is constricted above the ovary.

→ Calyx - 5 lobed, segments generally narrowed and pointed Petaloid, campanulate, gamosepalous. Imbricate, valvate or quincuncial aestivation

→ Corolla - 5 petals mostly sympetalous or free often deeply 5 lobed, valvate or imbricate, inserted on the calyx tube, campanulate to rotate or salverform

→ Androecium - stamens usually 5, inserted at various levels on the hypanthium, extrorse anthers, anthers often connate and their cells straight or variously curved or twisted. Stamens alternating with Petals. Anthers dehisce by longitudinal slits

→ Gynoecium — Tricarpellary (3 carpels), Syncarpous, Ovary inferior, usually unilocular (one celled), 3 Parietal placentas often meeting and filling up the cavity of the ovary or ultimately 3 celled. Placentation appears to be axile but not, the ovules are not born in the centre, because of the fusion and curving back of the placentas so, on reaching the carpellary walls the placentas fork into two and bear the ovules. Ovules - numerous, anatropous, style short, stigma 3 to 5 usually forked

→ Fruit - a succulent berry with hard rind, commonly known as pepo. dehiscent often explosively or more commonly indehiscent
→ Seeds - many, flattened exalbuminous, with straight embryo having large cotyledons, the seed coat consists of many layers, oily contents present
→ Pollination is entomophilous.

Families of Dicotyledons

Genera included under Cucurbitaceae

- Abobra
- Acanthosicyos
- Actinostemma
- Alsomitra
- Ampelosycios
- Anacaona
- Apatzingania
- Apodanthera
- Bambekea
- Benincasa
- Biswarea
- Bolbostemma
- Brandegea
- Bryonia
- Calycophysum
- Cayaponia
- Cephalopentandra
- Ceratosanthes
- Chalema
- Cionosicyos
- Citrullus
- Coccinia
- Cogniauxia
- Corallocarpus
- Cremastopus
- Ctenolepis
- Cucumella
- Cucumeropsis
- Cucumis
- Cucurbita
- Cucurbitella
- Cyclanthera
- Dactyliandra
- Dendrosicyos
- Dicoelospermum
- Dieterlea
- Diplocyclos
- Doyerea
- Ecballium
- Echinocystis
- Echinopepon
- Edgaria
- Elateriopsis
- Eureiandra
- Fevillea
- Gerrardanthus
- Gomphogyne
- Gurania
- Guraniopsis
- Gymnopetalum
- Gynostemma
- Halosicyos
- Hanburia
- Helmontia
- Hemsleya
- Herpetospermum
- Hodgsonia
- Ibervillea
- Indofevillea
- Kedrostis
- Lagenaria
- Lemurosicyos
- Luffa
- Marah
- Melancium
- Melothria
- Melothrianthus
- Microsechium
- Momordica
- Muellerargia
- Mukia
- Myrmecosicyos
- Neoalsomitra

- → Nothoalsomitra
- → Odosicyos
- → Oreosyce
- → Parasicyos
- → Penelopeia
- → Peponium
- → Peponopsis
- → Polyclathra
- → Posadaea
- → Praecitrullus
- → Pseudocyclanthera
- → Pseudosicydium
- → Psiguria
- → Pteropepon
- → Pterosicyos
- → Raphidiocystis
- → Ruthalicia
- → Rytidostylis
- → Schizocarpum
- → Schizopepon
- → Sechiopsis
- → Sechium
- → Selysia
- → Seyrigia
- → Sicana
- → Sicydium
- → Sicyos
- → Sicyosperma
- → Siolmatra
- → Siraitia
- → Solena
- → Tecunumania
- → Telfairia
- → Thladiantha
- → Trichosanthes
- → Tricyclandra
- → Trochomeria
- → Trochomeriopsis
- → Tumacoca
- → Vaseyanthus
- → Wilbrandia
- → Xerosicyos
- → Zanonia
- → Zehneria
- → Zombitsia
- → Zygosicyos.

Chapter - 79
Begoniaceae

Classification (Bentham and Hooker)

Phanerogams
Dicotyledons
Polypetalae
Calyciflorae
Passiflorales
Begoniaceae

General characters

→ Mostly herbs, shrubs or lianas; plants mostly succulent.

→ Leaves - alternate, spiral or distichous; flat, herbaceous or fleshy, petiolate, non sheathing, simple or compound; lamina usually entire and rarely dissected; stipulate.

→ Plants - Monoecious; the first inflorescence axes usually ending in male flowers; the last and sometimes the penultimate ones in females. Female flowers with or without staminodes. Gynoecium is absent in male flowers.

→ Inflorescence - cymose.

→ Flowers - irregular, cyclic
→ Perianth - petaline; 2 or 4 or 5 or 10 (rarely); free or rarely joined, 1 whorled rarely 2; white/cream/orange/red/pink. Corolla - 2 or 4 in 1 or 2 whorled usually polypetalous rarely gamopetalous; imbricate or valvate
→ Androecium 4-many; free of one another or joined (Monadelphous), 2-5 whorled. Anthers dehisce via pores or longitudinal slits;
→ Gynoecium - 2-6 carpelled, syncarpous, inferior. Placentation axile. Ovules 15-50 per locule, anatropous
→ Fruit - usually non fleshy, dehiscent, capsule or rarely fleshy, non-dehiscent berry
→ Seeds - non endospermic

Genera included under Begoniaceae

- Begonia
- Hillebrandia
- Symbegonia
- Begoniella
- Semibegoniella

Chapter – 80
Datisceae

Classification (Bentham and Hooker)
- Phanerogams
- Dicotyledons
- Polypetalae
- Calyciflorae
- Passiflorales
- Datisceae (Datiscaceae)

General characters
→ Tall, glabrous herbs
→ Leaves – alternate, petiolate, non-sheathing, compound, pinnate
→ Plants dioecious. Pollination anemophilous. Gynoecium is absent in male flowers
→ Inflorescence – fascicles. (Crowded fascicles on long, leafy branches)
→ Perianth – sepaline; calyx 3-6 (male flowers); 3-8 (female flowers); polysepalous, unequal, persistent
→ Androecium 3-5 (hermaphrodite flowers), 8-25 (male flowers); filantherous; Anthers dorsifixed; dehiscing via longitudinal slits;

extrorse
- Gynoecium 3-5 carpelled, syncarpous, synovarious (open at the apex); inferior; ovary 1 locular. Styles 3-5 (each deeply bifid). Stigma – dry type; Placentation – parietal; ovules many in single cavity, anatropous.
- Fruit – non-fleshy, dehiscent, membranous capsule
- Seeds – non endospermic with oily cotyledons

Genera included under Daticeae
- Dastica

Sub class: polypetalae

Series: calyciflorae

Order: ficoidales

Families:

Cactaceae

Ficoideae

Chapter - 81
Cactaceae

Classification (Bentham and Hooker)

    Phanerogams
    Dicotyledons
    Polypetalae
    Calyciflorae
    Ficoidales
    Cactaceae

General characters:

→ Succulent plants showing a varied and remarkable adaptations to xerophytic mode of life; range from small plants to large columnar trees.

→ Stems - fleshy and of various shapes. Often enlarged and cylindrical, flattened or fluted. Stems are frequently constricted and jointed. They are columnar, conical or subglobose angular, very often ribbed and warty. They are simple or branched. Sometimes they are rope like and flat like leaves. The stem of Pereskia show normal dicotyledonous growth with much branched and woody nature. Leaves are broad and fleshy. In Opuntia, the narrow leaves are found, but they fall off early. Leaves in majority of genera are small or underdeveloped.

In all the genera except Pereskia, assimilation

of food takes placed by stem; the tubercles are found to be developed upon the stem. Spines are borne upon the tubercule. Spine bearing area at the tip of the tubercule or along the ridges or edges of the columnar or flattened stem is called areole. Tubercles unite in vertical rows to form ribs. The spines correspond to the leaves.

→ Inflorescence – Usually the flowers are solitary and develop in the axils or at the end of the tubercles.

→ Flower – large, showy, bisexual, actinomorphic or zygomorphic (some case), epigynous. Zygomorphic is due to curving of the perianth, stamens and pistils.

→ Perianth not clearly distinguished into calyx and corolla. Number of floral leaves is large and the floral leaves are united below into a long tube. In some (Opuntia), perianth leaves are arranged spirally. The lower floral leaves resemble sepals being thick and green while the upper ones resemble the petals which are delicate and white or coloured.

→ Androecium consists of indefinite number of stamens which arise from perianth tube. They are arranged either spirally or in groups. Very often a ring of short stamens surround the throat of the tube. Stamens are epiphyllous. Anthers are basi or dorsifixed and two celled, dehisce by longitudinal slits. Pollens are very small, yellow, spherical, smooth and with three small germ pores.

→ Gynoecium usually consists of 4-α carpels that are united (syncarpous). Ovary - inferior, unilocular, more or less sunken in the floral axis, Parietal placentation. Parietal placentas being arranged on the interior of the ovary wall. Style is simple, Ovules are many each with two integuments.

→ Fruit - fleshy, one celled berry with numerous seeds. Fruit is dry in *Echinocactus*. The fruit is generally sweet and tasty.

→ Seeds are endospermic or non-endospermic with straight or curved embryo.

Genera included under Cactaceae

- Acanthocalycium
- Acanthocereus
- Aporocactus
- Ariocarpus
- Armatocereus
- Arrojadoa
- Arthrocereus
- Astrophytum
- Austrocactus
- Aztekium
- Bergerocactus
- Blossfeldia
- Brachycereus
- Brasilicereus
- Browningia
- Calymmanthium
- Carnegiea
- Cephalocereus
- Cereus
- Cipocereus
- Cleistocactus
- Coleocephalocereus
- Copiapoa
- Corryocactus
- Coryphantha
- Denmoza
- Discocactus
- Disocactus
- Echinocactus
- Echinocereus
- Echinopsis
- Epiphyllum
- Epithelantha
- Eriosyce
- Escobaria
- Espostoa
- Espostoopsis
- Eulychnia
- Facheiroa
- Ferocactus
- Frailea
- Gymnocalycium
- Haageocereus
- Harrisia
- Hatiora
- Heliocereus
- Hylocereus
- Jasminocereus
- Leocereus
- Lepismium
- Leptocereus
- Leuchtenbergia
- Lophophora
- Maihuenia
- Mammillaria
- Melocactus
- Micranthocereus
- Mila
- Myrtillocactus
- Neolloydia
- Neoporteria
- Neoraimondia
- Neowerdermannia
- Obregonia
- Opuntia
- Oreocereus
- Pachycereus
- Parodia
- Pediocactus
- Pelecyphora
- Peniocereus
- Pereskia

- → Pereskeopsis
- → Pilosocereus
- → Pseudorhipsales
- → Pterocactus
- → Rebutia
- → Rhipsalis
- → Samaipaticereus
- → Schlumbergera
- → Sclerocactus
- → Selenicereus
  - → Stenocactus
  - → Stenocereus
  - → Stephanocereus
- → Stetsonia
- → Strombocactus
- → Tacinga
- → Thelocactus
- → Uebelmannia
- → Weberbauerocereus
- → Weberocereus
- → Zygocactus

Genera included under Ficoideae:

→ Reaumuria
→ Nitraria
→ Sesuvium
→ Aizoon
→ Glinus
→ Oxygia
→ Mesembryanthemum
→ Tetragonia

Chapter - 82
Systematic Position
- Phanerogams
- Dicotyledons
- Polypetalae
- Calyciflorae
- Ficoidales
- Ficoideae

General characters

→ Herbs
→ Prostrate, much branched, growing in patches
→ Leaves - spathulate, apparently verticellate
→ Flowers on long axillary pedicels, clustered into a sort of umbel
→ Sepals 5, white inside
→ Petals absent
→ Stamens - 3
→ Styles 3, Pod - 3 celled, 3-valved, loculicidal
→ Many seeds.

Sub class: Polypetalae

Series: Calyciflorae

Order: Umbellales

Families:

Umbelliferae

Araliaceae

Cornaceae

Chapter - 83
Apiaceae (Umbelliferae)

Classification (Bentham and Hooker)

- Phanerogams
- Dicotyledons
- Polypetalae
- Calyciflorae
- Umbellales
- Apiaceae

General characters

→ Majority are annual, biennial or perennial herbs. Sometimes shrubs and undershrubs are present. *Pseudocaruny* - climbing plant.

→ Branched tap root system

→ Stem - usually erect and herbaceous. The stem is usually green and becomes pinkish as the plants mature. The stems are very often ribbed and angled.

→ Leaves - alternate, exstipulate amplexicaul and much dissected, rarely opposite; pinnately compound rarely simple. Species *Eryngium* and *Acephylla* possess the leaves with parallel venation and sheathing bases like monocots. Oil glands are present in all the aerial parts of the plants.

→ Inflorescence - umbel. May be simple or compound.

- Flower – hermaphrodite rarely unisexual; usually actinomorphic rarely zygomorphic; pedicellate, bracteate, complete and epigynous.
- Calyx consists of 5 sepals which are very minute. Odd sepal being posterior. Many cases calyx is absent.
- Corolla consists of 5 petals, polypetalous, usually white or yellow in colour. In many cases 2 of the petals are larger in size than the other 3. Tips of the petals are usually reflexed. Sometime petals are bifid. Aestivation – valvate or imbricate.
- Androecium consists of 5 stamens alternate to the petals. Stamens arise from an epigynous disc. Anthers – bilobed, introrse, dorsifixed; filaments free. Anthers split by longitudinal slits.
- Gynoecium – bicarpellary, syncarpous, ovary inferior, bilocular, each locule contains a single pendulous, anatropous ovule. Axile placentation. On the top of the ovary a nectar disc is found which surrounds the two capitate stigmas.
- Fruit – cremocarp.
- Seed – endospermic with minute embryo.
- Pollination is entomophilous.

Genera included under Apiaceae

- Aciphylla
- Acronema
- Actinolema
- Actinotus
- Adenosciadium
- Aegopodium
- Aethusa
- Afranium
- Afrocarum
- Afroligusticum
- Afrosison
- Agasyllis
- Agrocharis
- Ainsworthia
- Albovia
- Alepidea
- Aletes
- Alococarpum
- Ammi
- Ammodaucus
- Ammoides
- Ammoselinum
- Anethum
- Angelica
- Anginon
- Angoseseli
- Aniopoda
- Anisosciadium
- Anisotome
- Annesorhiza
- Anthriscus
- Aphanopleura
- Apiastrum
- Apium
- Apodicarpum
- Arctopus
- Arcuatopterus
- Arracia
- Artedia
- Ascladium
- Asteriscium
- Astomaea
- Astrantia
- Astrodaucus
- Astydamia
- Athamanta
- Aulacospermum
- Autumnalia
- Azilia
- Azorella
- Berula
- Bifora
- Bilacunaria
- Bolax
- Bonannia
- Bowlesia
- Bunium
- Bupleurum
- Cachyrs
- Calyptosciandium
- Capnophyllum
- Carlesia
- Caropsis
- Carum
- Caucalis
- Cenolophium
- Centella
- Cephalopodium
- Chaerophyllopsis
- Chaerophyllum
- Chaetosciadium
- Chamaele

Families of Dicotyledons

- → Chamaesciadium
- → Chamaesium
- → Chamarea
- → Changium
- → Chlaenosciadium
- → Choritaenia
- → Chuanminshen
- → Chymsydia
- → Cyclospermum
- → Cenolata
- → Cnideocarpa
- → Cnidium
- → Coaxana
- → Conioselinum
- → Conium
- → Conopodium
- → Coriandrum
- → Costia
- → Cortiella
- → Cotopaxia
- → Coulterophytum
- → Crenosciadium
- → Crithmum
- → Cryptotaenia

- → Cuminum
- → Cyathoselinum
- → Cyclorhiza
- → Cymbocarpum
- → Cymopterus
- → Cynosciadium
- → Dactylaea
- → Dasispermum
- → Daucosma
- → Daucus
- → Dethawia
- → Deverra
- → Dichisciadium
- → Dickinsia
- → Dicyclophora
- → Dimorphosciadium
- → Diplaspis
- → Diplolophium
- → Diplotaenia
- → Dipposis
- → Domeykoa
- → Donnellsmithia
- → Dorema

- → Dracosciadium
- → Drusa
- → Ducrosia
- → Dystaenia
- → Echinophora
- → Elaeoselinum
- → Eleutherospermum
- → Enantiophylla
- → Endressia
- → Eremocharis
- → Eremodaucus
- → Ergocarpon
- → Erigenia
- → Eriocycla
- → Eriosynaphe
- → Eryngium
- → Erythroselinum
- → Eurytaenia
- → Exoacantha
- → Ezosciadium
- → Falcaria
- → Fergania
- → Ferula
- → Ferulago
- → Foeniculum

- → Frommia
- → Fooriepa
- → Fuernrohria
- → Galagania
- → Geocaryum
- → Gingidia
- → Glaucosciadium
- → Glehnia
- → Glia
- → Glochidotheca
- → Gongylosciadium
- → Grafia
- → Grammosciadium
- → Hacquetia
- → Halosciatrum
- → Haplosciadium
- → Haplosphaera
- → Harbouria
- → Harrysmithia
- → Hausknechtia
- → Hellenocarum
- → Heptaptera
- → Heracleum
- → Hermas
- → Heteromorpha
- → Hladnekia
- → Hohenackeria
- → Homalocarpus
- → Homalosciadium
- → Horstrissea
- → Huanaca
- → Hyalolaena
- → Hydrocotyle
- → Itasina
- → Johrenia
- → Kadenia
- → Kafirnigania
- → Kalakia
- → Kandaharia
- → Karatavia
- → Karnataka
- → Kedarnatha
- → Keraymonia
- → Kitagawia
- → Klotzschia
- → Komarovia
- → Korovinia
- → Korshinskia
- → Kosopoljanskia
- → Koslovia
- → Krasnovia
- → Krubera
- → Kundmannia
- → Ladyginia
- → Lagoecia
- → Laretia
- → Laser
- → Laserpitium
- → Lecokia
- → Ledebourella
- → Lefebvrea
- → Lereschia
- → Levisticum
- → Lichtensteinia
- → Lignocarpa
- → Ligusticopsis
- → Ligusticum
- → Lilaeopsis
- → Limnosciadium
- → Lisaea
- → Lomatium
- → Lomatopodium

→ Magadania
→ Magydaris
→ Malabaila
→ Mandenovia
→ Marlothiella
→ Mastigosciadium
→ Mathiasella
→ Mediasia
→ Meeboldia
→ Melanosciadium
→ Melanoselinum
→ Merwiopsis
→ Meum
→ Micropleura
→ Microsciadeum
→ Mogoltavia
→ Molopospermum
→ Monizia
→ Mulinum
→ Muretia
→ Musineon
→ Myrrhidendron
→ Myrrhis
→ Myrrhoides

→ Naufraga
→ Neocryptodiscus
→ Neogoezia
→ Neonelsonia
→ Neoparrya
→ Neoplatytaenia
→ Neosciadium
→ Niphogeton
→ Niranathamnos
→ Nothosmyrnium
→ Notiosciadium
→ Notopterygium
→ Oedibasis
→ Oenanthe
→ Oligocladus
→ Oliveria
→ Olymposciadium
→ Oropanax
→ Oreocome
→ Oreomyrrhis
→ Oreonana
→ Oreoschimperella
→ Oreoxis
→ Orlaya

→ Ormopterum
→ Ormosciadeum
→ Orogenia
→ Oschatzia
→ Osmorhiza
→ Ottoa
→ Oxypolis
→ Pachyctenium
→ Pachypleurum
→ Palimbia
→ Paralegusticum
→ Paraselinum
→ Parasilaus
→ Pastinaca
→ Pastinacopsis
→ Pauleta
→ Pedinopetalum
→ Pentapeltis
→ Perideridia
→ Perissocoelium
→ Petagnaea
→ Petroedmondia
→ Petroselinum
→ Peucedanum

- → Phellolophium
- → Phlojodicarpus
- → Phlyctodocarpa
- → Physospermopsis
- → Physospermum
- → Physotrichia
- → Pelopleura
- → Pimpinella
- → Pinda
- → Platysace
- → Pleurospermopsis
- → Pleurospermum
- → Podistera
- → Polemannia
- → Polemanniopsis
- → Polylophium
- → Polytaenia
- → Polyzygus
- → Portenschlagiella
- → Proa
- → Prangos
- → Prionosciadium
- → Psammogeton
- → Pseudocarum
- → Pseudorlaya
- → Pseudoselinum
- → Pternopetalum
- → Pterygopleurum
- → Ptilimnium
- → Ptychotis
- → Pycnocycla
- → Pyramidoptera
- → Registaniella
- → Rhabdosciadium
- → Rhodosciadium
- → Rhopalosciadium
- → Rhysopterus
- → Ridolfia
- → Rouya
- → Rumia
- → Rutheopsis
- → Sajanella
- → Sanicula
- → Saposhnikovia
- → Scaligeria
- → Scandia
- → Scandix
- → Schizeilema
- → Schoenolaena
- → Schrenkia
- → Schtschurowskia
- → Schulzia
- → Schumannia
- → Sclerochorton
- → Sclerotiaria
- → Scrithacola
- → Selinum
- → Semenovia
- → Seseli
- → Seselopsis
- → Shoshonea
- → Silaum
- → Sinocarum
- → Sinodielsia
- → Sinolimpaichtia
- → Sison
- → Sium
- → Smyrniopsis
- → Smyrnium
- → Sonderina
- → Soranthus

- Spananthe
- Spermolepis
- Sphaenolobium
- Sphaerosciadium
- Sphallerocarpus
- Sphenocarpus
- Sphenosciadium
- Spongiosyndesmus
- Spuriodaucus
- Spuriopimpinella
- Stefanoffia
- Steganotaenia
- Stenocoelium
- Stenosemis
- Stenotaenia
- Stewartiella
- Strobrax
- Symphyoloma
- Synelcosciadium
- Szovitsia
- Taenidia
- Taniamschjania
- Tauschia
- Tetrataenium
- Thamnosciadium
- Thapsia
- Thaspium
- Thecocarpus
- Tilingia
- Tinguarra
- Todaroa
- Tongoloa
- Tordyliopsis
- Tordylium
- Torilis
- Tornabenea
- Trachydium
- Trachymene
- Trachysciadium
- Trachyspermum
- Transcaucasia
- Trepocarpus
- Tricholaser
- Trigonosciadium
- Trinia
- Trochiscanthes
- Turgenia
- Uldinia
- Vanasushava
- Vicatia
- Xanthogalum
- Xanthosia
- Xatardia
- Yabea
- Zeravschania
- Zizia
- Zosima

Chapter - 84
Araliaceae

Classification (Bentham and Hooker)

- Phanerogams
- Dicotyledons
- Polypetalae
- Calyciflorae
- Umbellales
- Araliaceae

General characters

→ Mostly trees or shrubs or leanas or herbs.
→ Leaves - alternate or opposite or whorled, spiral or distichous; petiolate; leaf sheaths with free margins; stipulate or exstipulate.
→ Plants hermaphrodite or monoecious.
→ Flowers in inflorescence - heads, spikes, umbels
→ Flowers - rarely calyptrate, pentamerous, cyclic.
→ Perianth with distinct calyx and corolla or petaline.
→ Calyx - 3-12 sepals, 1 whorled, poly or gamo sepalous, entire / lobulate.

→ Corolla - 3-13 petals; 1 whorled; commonly alternating with calyx, polypetalous or partially gamopetalous; imbricate or valvate.

→ Androecium 3-12 or 10-100 stamens, isomerous with perianth or diplostemonous; filantherous. Anthers - dorsifixed; dehiscing via longitudinal slits; introrse.

→ Gynoecium - 1 to 100 carpelled; syncarpous, partly inferior to superior; ovule -1; ovary - 1 to 100 locular. Epigynous disk present; Style - free or joined, apical; stigma present

→ Fruit - fleshy or non-fleshy; indehiscent or schizocarp (berry or drupe).

→ Seeds - endospermic

Genera included under Araliaceae

- Anakasia
- Apiopetalum
- Aralia
- Arthrophyllum
- Astrotricha
- Boninofatsia
- Brassaiopsis
- Cephalaralia
- Cussonia
- Delarbrea
- Dendropanax
- Eleutherococcus
- Fatsia
- Gamblea
- Gastonia
- Harmsiopanax
- Hedera
- Heteropanax
- Hunaniopanax
- Kalopanax
- Mackinlaya
- Macropanax
- Megalopanax
- Merrilliopanax
- Meryta
- Motherwellia
- Munroiodendron
- Myodocarpus
- Neopanax
- Oplopanax
- Oreopanax
- Osmoxylon
- Panax
- Pentapanax
- Polyscias
- Pseudopanax
- Pseudosciadium
- Raukaua
- Reynoldsia
- Schefflera
- Sciadodendron
- Seemannaralia
- Sinopanax
- Stilbocarpa
- Tetrapanax
- Tetraplasandra
- Trevesia
- Woodburnia

Chapter - 85
Cornaceae

Classification (Bentham and Hooker)

  Phanerogams
  Dicotyledons
  Polypetalae
  Calyciflorae
  Umbellales
  Cornaceae

General characters

→ Trees and shrubs or herbs; plants (herbs) — Perennial, rhizomatous.

→ Leaves — persistent or deciduous, opposite or alternate; herbaceous or leathery; lamina — entire or dissected; exstipulate.

→ Plants usually hermaphrodite, rarely dioecious.

→ Inflorescence — cymes, heads, corymbs, umbels.

→ Flowers — bracteate, 4-5 merous, cyclic.

→ Perianth with distinct calyx and corolla (or) sepaline;

→ Calyx 4-7 in 1 whorled, gamosepalous; entire or lobate.

→ Corolla when present — 4-5 in 1 whorled, polypetalous, valvate, regular.

→ Androecium - 4-5 ; 1 whorled ; oppositisepalous. Anthers - introrse ; dehiscing via longitudinal slits

→ Gynoecium 2-4 carpelled ; syncarpous, inferior ; ovary 1 locular ; epigynous disk present ; styles 1 or 2-4, free or partially joined, apical ; stigma present ; placentation usually axile or apical when unilocular - parietal ; ovules 1 per locule

→ Fruit - fleshy to non-fleshy, indehiscent, drupe or rarely berry.

→ Seeds endospermic

Genera included under Cornaceae

→ Cornus

Sub class: Gamopetalae

Series: Inferae

Order: Rubiales

Families:

Caprifoliaceae

Rubiaceae

Chapter - 86
Caprifoliaceae

Classification (Bentham and Hooker)

- Phanerogams
- Dicotyledons
- Gamopetalae
- Inferae
- Rubiales
- Caprifoliaceae

General characters

→ Shrubs or small trees or rarely herbs or lianas

→ Leaves - persistent or deciduous; usually opposite or whorled; mostly herbaceous or leathery; petiolate; simple; lamina dissected. Stipulate or exstipulate

→ Plants hermaphrodite

→ Perianth with distinct calyx and corolla; Calyx consists of 2-5 sepals in 1 whorled; usually gamosepalous; imbricate

→ Corolla consists of 4-5 petals in 1 whorl; gamopetalous; imbricate; campanulate or funnel shaped or tubular; white/red/yellow;

Pink/Purple

→ Androecium 2-5 stamens, epipetalous, 1 whorled; didynamous; isomerous with the perianth; oppositisepalous; Anthers separate from one another; dorsifixed, introrse; dehisce via longitudinal slits.

→ Gynoecium 2-8 carpelled; Carpels reduced in number relative to that of perianth; syncarpous, inferior; ovary 2-8 locular; epigynous disk may be present; Styles-1, apical; Stigma- 1, usually capitate rarely truncate. Placentation axile to apical; ovules 1-50 per locule;

→ Fruit - fleshy or non-fleshy; dehiscent or indehiscent; capsule or achene like or berry.

→ Seeds - endospermic.

Genera included under caprifoliaceae
- Abelia
- Diervilla
- Dipelta
- Heptacodium
- Kolkwitzia
- Leycesteria
- Linnaea
- Lonicera
- Symphoricarpos
- Triosteum
- Weigela
- Zabelia

Chapter - 87
Rubiaceae

Classification (Bentham and Hooker)

- Phanerogams
- Dicotyledons
- Gamopetalae
- Inferae
- Rubrales
- Rubiaceae

General characters:

→ Majority - trees or shrubs; sometimes - herbaceous, climber. *Myrmecodia* possesses a tuber like stem developed from the swollen hypocotyl. This tuber like stem contains a number of communicating galleries which harbour the ants. Such type of plants are known as myrmecophilous.

→ Leaves - simple, stipulate, opposite decussate, entire or rarely toothed. Stipules may be inter or intra petiolar. Sometimes stipules unite to each other and then to the petiole, forming a sheath like structure around the stem.

→ Inflorescence - cymose (dichasial or panicled cyme). Sometimes small flowered dichasia are aggregated into dense globose heads. *Ixora* - corymbose cymes. *Oldenlandia* - Panicled cymes.

→ Flowers - hermaphrodite, actinomorphic and either tetramerous or pentamerous, complete, epigynous.

- Coprosma – flowers unisexual by reduction
- → Henriquezia – flowers are zygomorphic
- → Calyx consists of 5 or 4 sepals, gamosepalous. Sometimes one or rarely more sepals become leaf like and white, yellow or variously coloured. Development of such leaf like sepal increases the attractiveness of the inflorescence and attraction of the insects. Aestivation – Valvate.
- → Corolla consists of 5 or 4 Petals, gamopetalous; Corolla is tubular, campanulate or rotate. Aestivation – contorted or valvate
- → Androecium consists of 5 or 4 stamens which are epipetalous (ie) they are inserted near the throat of the corolla tube. Stamens alternate with the petals. Anthers – bicelled, introrse, dehisce by longitudinal slits or sometimes by apical pores.
- → Gynoecium – 2 carpels, Syncarpous. Generally ovary is inferior rarely half inferior or superior. Generally bilocular rarely unilocular; Usually Placentation is axile, rarely parietal. Style simple with globose stigma
- → Fruit – drupe, capsule or berry. May be dry or fleshy
- → Seeds – small and more or less winged, endospermic.
- → Pollination – entomophilous.

Families of Dicotyledons

Genera included under Rubiaceae

- Acranthera
- Acrobotrys
- Acunaeanthus
- Adinauclea
- Agathisanthemum
- Aidia
- Aitchisonia
- Alberta
- Aleisanthia
- Alibertia
- Allaeophania
- Alleizetella
- Allenanthus
- Alseis
- Amaioua
- Amaracarpus
- Amphiasma
- Amphidasya
- Ancylanthos
- Anomanthodia
- Antherostele
- Anthorrhiza
- Anthospermum
- Antirhea
- Aoranthe
- Aphaenandra
- Aphanocarpus
- Appunettia
- Appunia
- Arctophyllum
- Argocoffeopsis
- Argostemma
- Aseminantha
- Asperula
- Astiella
- Atractocarpus
- Atractogyne
- Augusta
- Aulacocalyx
- Badusa
- Balmea
- Balaprune
- Bathysa
- Batopedina
- Belonophora
- Benkara
- Benzonia
- Berghesia
- Bertiera
- Bikkia
- Blandibractea
- Blepharidium
- Bobea
- Boholia
- Borojoa
- Bothriospora
- Botryarrhena
- Bouvardia
- Brachytome
- Bradea
- Brenania
- Breonadia
- Burchellia
- Burttdavya
- Byrsophyllum
- Caelospermum
- Calanda
- Callipeltis
- Calochone
- Captaincookia
- Carpacoce
- Carphalea
- Carterella
- Casasia
- Catesbaea
- Catunaregam
- Cephaelis

- → Cephalodendron
- → Ceratopyxis
- → Ceriscoides
- → Ceuthocarpus
- → Chaetostachydium
- → Chalepophyllum
- → Chamaepentas
- → Chapelieria
- → Chassalia
- → Chazaliella
- → Chimarrhis
- → Chiococca
- → Chione
- → Chlorochorion
- → Chomelia
- → Choulettia
- → Chytropsia
- → Cigarilla
- → Cinchona
- → Cladoceras
- → Clarkella
- → Coccochondra
- → Coccocypselum
- → Coddia
- → Coelopyrena
- → Coffea
- → Coleactina
- → Colletoecima
- → Commitheca
- → Condaminea
- → Conostomium
- → Coprosma
- → Coptophyllum
- → Coptosapelta
- → Coptosperma
- → Corynanthe
- → Coryphothamnus
- → Cosmibuena
- → Cosmocalyx
- → Coursiana
- → Coussarea
- → Coutaportla
- → Coutarea
- → Cowiea
- → Craterispermum
- → Cremaspora
- → Cremocarpon
- → Crobylanthe
- → Crocyllis
- → Crossopteryx
- → Crucianella
- → Cruciata
- → Cruckshanksia
- → Crusea
- → Cuatrecasasiodendron
- → Cubanola
- → Cuviera
- → Cyclophyllum
- → Damnacanthus
- → Danais
- → Deccania
- → Declieuxia
- → Dendrosipanea
- → Dentella
- → Deppea
- → Diacrodon
- → Dialypetalum
- → Dibrachyonostylus
- → Dichilanthe
- → Dictyandra
- → Didymaea
- → Didymochlamys
- → Didymopogon
- → Didymosalpynx
- → Diodella
- → Diodia
- → Dioecrescis
- → Dioicodendron
- → Diplospora
- → Discospermum

- → Diyaminanclea
- → Dolichodelphys
- → Dolichololium
- → Dolichometra
- → Doricura
- → Ducidania
- → Dunnia
- → Duperrea
- → Durioia
- → Durringtonia
- → Ecpoma
- → Elzia
- → Elaeagia
- → Eleutheranthus
- → Emmenopterys
- → Enmerchiza
- → Empogona
- → Eosanthe
- → Eriosemopsis
- → Erithalis
- → Eonodea
- → Etericius
- → Euclinia
- → Exallage
- → Exostema
- → Fadogia
- → Fadogiella
- → Jagerlindia
- → Fararpea
- → Ferdinandusa
- → Feretia
- → Fergusonia
- → Fernelia
- → Flagenium
- → Hexanthera
- → Gaertnera
- → Gaillonia
- → Galeniera
- → Galium
- → Gallienia
- → Galopina
- → Ganiotopea
- → Gardenia
- → Gardeniopsis
- → Genipa
- → Gentingia
- → Geophela
- → Gillesprea
- → Gleasonia
- → Glionnetia
- → Glossostipula
- → Gomphocalyx
- → Gonzalagunia
- → Gouldia
- → Greenea
- → Greeniopsis
- → Guettarda
- → Gynochthodes
- → Gyrostipula
- → Halxoneuron
- → Haldina
- → Hallea
- → Hamelia
- → Hyatalla
- → Hedstromia
- → Hedyotis
- → Hedythyrsus
- → Heinsenia
- → Heinsia
- → Hekistocarpa
- → Heterophyllaea
- → Hillia
- → Himalrandia
- → Hindsia
- → Hintonia

→ Hippotis
→ Hitoa
→ Hodgkinsonia
→ Hoffmannia
→ Holstianthus
→ Homollea
→ Homolliella
→ Houstonia
→ Hutchinsonia
→ Hydnophytum
→ Hydrophylax
→ Hymenocnemis
→ Hymenocoleus
→ Hymenodictyon
→ Hyperacanthus
→ Hypobathrum
→ Hyptianthera
→ Ibetralia
→ Indopolysolenia
→ Isertia
→ Isidorea
→ Ixora
→ Jackiopsis
→ Janotia
→ Jaubertia
→ Joosia

→ Jovetia
→ Kailarsenia
→ Knoxia
→ Kochummenia
→ Kohautia
→ Kraussia
→ Kutchubaea
→ Labenbergia
→ Lagynias
→ Lamprothamnus
→ Lasianthus
→ Lathraeocarpa
→ Lecananthus
→ Lecanosperma
→ Lecariocalyx
→ Lelya
→ Lemyrea
→ Lepidostoma
→ Leptactina
→ Leptodermis
→ Leptomischus
→ Leptoscela
→ Leptostigma
→ Lerchea
→ Leucocodon
→ Leucolophus

→ Limnosipanea
→ Lindenia
→ Litosanthes
→ Lucinaea
→ Luculia
→ Lucya
→ Ludekia
→ Macbridenia
→ Machaonia
→ Macrocnemum
→ Macrosphyra
→ Mantalania
→ Margaritopsis
→ Maschalocorymbus
→ Maguireocharis
→ Maguireothamnus
→ Malanea
→ Manettia
→ Manostachya
→ Maschalodesme
→ Massularia
→ Mastixiodendron
→ Mazaea
→ Melanopsidium
→ Mericarpaea
→ Merumea
→ Metadina

- → Heyna
- → Micrasepalum
- → Microphysa
- → Mitchella
- → Metracarpus
- → Mitrasacmopsis
- → Metriostigma
- → Molopanthera
- → Monosalpinx
- → Montamans
- → Morelia
- → Moriering
- → Morinda
- → Morindopsis
- → Motleyia
- → Mouretia
- → Multidentia
- → Mussaenda
- → Mussaendopsis
- → Mycetia
- → Myonima
- → Myrmecodia
- → Myrmeconauclea
- → Myrmephytum
- → Naletonia
- → Nargedia
- → Neanotis
- → Neblinathamnus
- → Nematostylis
- → Nenax
- → Neobertiera
- → Neoblakea
- → Neobreonia
- → Neofranciella
- → Neohymenopogon
- → Neolamarckia
- → Neolaugeria
- → Neopentanisia
- → Nernstia
- → Nertera
- → Nesohedyotis
- → Neurocalyx
- → Nichallea
- → Nodocarpaea
- → Nonatelia
- → Normandia
- → Nostolachma
- → Ochreinauclea
- → Octotropis
- → Oldenlandia
- → Oldenlandiopsis
- → Oligocodon
- → Omiltemia
- → Opercularia
- → Ophiorrhiza
- → Ophiococcus
- → Oreeandra
- → Oreopolus
- → Osa
- → Otiophora
- → Otocalyx
- → Otomeria
- → Ottoschmidtia
- → Oxyanthus
- → Pachystigma
- → Pachystylus
- → Paederia
- → Paganyea
- → Paganiopsis
- → Palicourea
- → Pamplethantha
- → Paracephalis
- → Parachimarrhis
- → Paracorynanthe
- → Paragenipa
- → Paraknoxia
- → Parapentas
- → Paratriana
- → Pauridiantha

→ Pausinystalia
→ Pavetta
→ Pelagodendron
→ Pentagonia
→ Pentaloncha
→ Pentanisia
→ Pentanopsis
→ Pentas
→ Pentodon
→ Peponidium
→ Perakanthus
→ Perama
→ Peratanthe
→ Peripeplus
→ Pertusadina
→ Petagomoa
→ Petitiocodon
→ Phellocalyx
→ Phialanthus
→ Philopis
→ Phuopsis
→ Phyllacanthus
→ Phyllis
→ Phyllocrater
→ Phyllomelia

→ Phyllohydrax
→ Picardaea
→ Pimentelia
→ Pinarophyllon
→ Pinckneya
→ Pittoniotis
→ Placocarpa
→ Placopoda
→ Plectronia
→ Plectroniella
→ Pleiocarpidia
→ Plerocoryne
→ Plerocraterium
→ Plocama
→ Poecilocalyx
→ Pogonolobus
→ Pogonopus
→ Polysphaeria
→ Polyura
→ Pomax
→ Porterandia
→ Portlandia
→ Posoqueria
→ Pouchetia
→ Praravinia

→ Pravinaria
→ Preussiodosa
→ Primatomeris
→ Proscephaleium
→ Psathura
→ Pseudaidia
→ Pseudogaillonia
→ Pseudogardenia
→ Pseudohamelia
→ Pseudomantalania
→ Pseudomussaenda
→ Pseudonesohedyotis
→ Pseudopyxis
→ Pseudosabicea
→ Psilanthus
→ Psychotria
→ Psydrax
→ Psylocarpus
→ Pteridocalyx
→ Pterogaillonia
→ Pubistylus
→ Putoria
→ Pygmaeothamnus
→ Pyragra
→ Pyrostria

- Rachicallis
- Ramosmania
- Randia
- Ranitebe
- Readea
- Remijia
- Rennellia
- Retiniphyllum
- Rhadinopus
- Raphiolura
- Rhipidantha
- Rhopalobrachium
- Rhyssocarpus
- Richardia
- Riquexuria
- Robynsia
- Roigella
- Ronabea
- Rondeletia
- Rothmannia
- Rubia
- Rudgea
- Rustia
- Rutidea
- Rytigynia
- Sabicea
- Sacosperma
- Saldinia
- Salzmannia
- Saprosma
- Sarcopygme
- Schachtia
- Schumatoclada
- Schizenterospermum
- Schizocalyx
- Schizocolea
- Schizomussaenda
- Schizostigma
- Schmidtottia
- Schradera
- Schumanniophyton
- Schwendenera
- Scolosanthus
- Scyphiphora
- Scyphochlamys
- Scyphostachys
- Sericanthe
- Serissa
- Shaferocharis
- Sherardia
- Sherbournia
- Siderobombyx
- Sicyensia
- Simira
- Sinodina
- Sipanea
- Sipaneopsis
- Siphonandrium
- Sommera
- Spathichlamys
- Spermacoce
- Spermadictyon
- Sphinctanthus
- Spiradiclis
- Squamellaria
- Stachyarrhena
- Stachyococcus
- Staelia
- Standleya
- Steenisia
- Stelechantha
- Stephanococcus
- Stevensia
- Steyermarkia
- Stichianthus
- Stilpnophyllum
- Stipularia
- Stomandra

- → Streblosa
- → Streblosiopsis
- → Strempelia
- → Striolaria
- → Strumpfia
- → Stylosiphonia
- → Subcoanthus
- → Sukunia
- → Sulitia
- → Synaptantha
- → Syringantha
- → Tamilnadia
- → Tammsia
- → Tapiphyllum
- → Tarenna
- → Tarennoidea
- → Temnocalyx
- → Temnopteryx
- → Tennantia
- → Tetralopha
- → Thecorchus
- → Thogsennia
- → Thyridocalyx
- → Timonius
- → Tobagoa
- → Tocoyena
- → Tortuella
- → Trailliaedoxa
- → Tresanthera
- → Trichostachys
- → Trigonopyren
- → Trukia
- → Tsiangia
- → Uragoga
- → Urophyllum
- → Valantia
- → Vangueria
- → Vangueriella
- → Vangueriopsis
- → Versteegia
- → Villaria
- → Virectaria
- → Warburgina
- → Warszewiczia
- → Wendlandia
- → Wernhamia
- → Wiasemskya
- → Wittmackanthus
- → Xanthophytum
- → Xantonnea
- → Xerococcus
- → Yutajea
- → Zuccarinia
- → Zygoon.

Sub class: Gamopetalae

Series: Inferae

Order: Asterales

Families:

Valerianeae

Dipsaceae

Calycereae

Compositae

Chapter - 88
Valerianeae

Classification (Bentham and Hooker)
Phanerogams
Dicotyledons
Gamopetalae
Inferae
Asterales
Valerianeae

General characters
→ Herbaceous, foliage often strong, disagreeable odor.
→ Ornamentals and some used in herbal medicine
→ Calyx composed of a tube and border, the border rolled inwards in flowers. Forms a feathery border (Pappus) in fruit
→ Corolla - 5 - united into a tube; the tube being sometimes pouched or spurred at the base; inserted in the throat of the calyx.
→ Stamens 1 or 3, rarely 8, inserted on the corolla tube (epipetalous)
→ Pistil - 1 to 3 carpels, united into a ovary; a

thread like style and 2-3 united or free stigmas.

→ Fruit – small and dry, crowned with the teeth of feathery border of the calyx; 3 celled with 1 cell only containing 1 hanging seed, indehiscent

→ Flowers – small, usually white, red or pink.

→ Inflorescence – cymose corymbs or heads.

→ Leaves – opposite and without stipules

Genera included under Valerianaceae

→ Centranthus
→ Nardostachys
→ Fedia
→ Patrinia
→ Plectritis
→ Valeriana
→ Valerianella.

Chapter-89
Dipsaceae

Classification (Bentham and Hooker)

Phanerogams
Dicotyledons
Gamopetalae
Inferae
Asterales
Dipsaceae

General characters

→ Anthers are free
→ Calyx - Distinct tubular calyx invested in a separate involucel (little involucre) of tiny bracts
→ Perennial; stout rootstock; hairy stem
→ Leaves - usually entire, oblong lance-shaped with toothed margins (leaves from root)
→ Leaves from stem are lobed sometimes almost pinnate
→ The flower-heads are borne on a long stout stalk
→ Involucral bracts broad and leaf like in 2 rows.

Genera included under Dipsaceae

- Acanthocalyx
- Dipsacus
- Knautia
- Scabiosa
- Succisa
- Succisella
- Morina
- Cephalaria
- Pterocephalus
- Pycnocomon
- Triplostegia

Families of Dicotyledons

Chapter-90
Calycereae (Calyceraceae)
(Bentham and Hooker)

Classification
- Phanerogams
- Dicotyledons
- Gamopetalae
- Inferae
- Asterales
- Calycereae

General characters

→ Perennial or annual herbs

→ Few or many branched stems that may be without hair or with soft silky hairs

→ Leaves - Rosette at the base of the stems or set alternately along the stem; exstipulate; simple but may be lobed. Margin - entire or toothed

→ Inflorescence - heads

→ Inflorescence are at the top of the stems or opposite leaves and may have a flowerstem

→ Each individual flowerhead is surrounded by an involucre, consisting of one or two rows of bracts that are often leaf like and not

usually merged
- Base of the flowerhead may be conical, convex or spheroidal
- Flowers – hermaphrodite or unisexual
- Petals fused to form a funnel shaped or cylinder shaped corolla which is split into 4 to 6 lobes at the top.
- Corolla persistent in fruit top
- 4 or 5 stamens alternate the corolla lobes. The lower thread of these filaments are fused with the corolla tube.
- Filaments carry nectaries; anthers stand upright
- Style – thread like without hairs; sticking out above the corolla tube; stigma – club shaped or split into 2
- Ovary – bicarpellary, 1 pendulous and anatropous ovule.
- Fruit – achene with persistent calyx which may contains spines
- Seeds – fleshy endosperm

Genera included under Calycereae
- Acicarpha
- Calycera
- Moschopsis
- Boopis
- Gamocarpha
- Nastanthus

Chapter - 9
Asteraceae (Compositae)

Classification (Bentham and Hooker)

Phanerogams
Dicotyledons
Gamopetalae
Inferae
Asterales
Compositae

General characters

→ herbs, shrubs very rarely trees and climbers
→ Branched tap root system. Sometimes tuberously thickened (Helianthus)
→ Leaves - alternate, sometimes opposite rarely whorled simple or compound; sometimes acicular (needle like) or reduced to scales in some xerophytes; exstipulate petiolate; surface glabrous or hairy; Margin entire or serrated; apex acute or obtuse; reticulate venation
→ Inflorescence - Racemose, head, capitulum with an involucre of bracts; rarely in spikes
→ Flower: Florets of a head may be hermaphrodite or unisexual or neutral (asexual); Pentamerous, actinomorphic or zygomorphic, epigynous.

There are 2 kinds of florets (i) Disc florets

(tubular flowers) (ii) Ray florets (ligulate flowers)
Arrangement of florets in a head are as follows:
a) All the ray and disc florets on a single head may be tubular and actinomorphic (*Ageratum*)
b) All the ray and disc florets in the head may be ligulate and zygomorphic (*Sonchus*)
c) Disc florets (central flowers) may be actinomorphic and tubular while the ray florets (marginal flowers) may be zygomorphic and ligulate or bilabiate

When all the flowers are identical in a head the condition is called homogamous; when unidentical — heterogamous flowers

[Disc floret]:
Sessile, regular, actinomorphic, complete, hermaphrodite, bracteate, epigynous

Calyx → very rudimentary or entirely absent. Sometimes calyx is modified into a large number of bristles or hair like structures called the Pappus. Pappus is persistent and act like a parachute in the fruit dispersal. When the bristles a few in number they are modified into barbs or spines which are persistent and may adhere to the body of animals and disseminated. Sometimes there is an epigynous five-lobed ring like structure

→ Corolla consists of 5 petals, gamopetalous (united) tubular, 5-lobed, actinomorphic, swollen near the base of style where nectary is present, valvate aestivation, variously coloured

→ Androecium consists of 5 stamens, epipetalous, alternate with corolla lobes. Anthers introrse, two celled, united laterally into a tube (syngenesious), dehiscing by longitudinal slits. Filaments free, connective generally prolonged above the anthers

→ Gynoecium - Bicarpellary, syncarpous, ovary inferior, unilocular, containing a single anatropous basal ovule; placentation-basal, stigmas 2, curved, the upper surface receptive.

Ray floret - Sessile, zygomorphic (irregular), ligulate or bracteate

Calyx - Sepals absent or represented by pappus
Corolla - Petals 5, gamopetalous, corolla may be bilabiate or ligulate
Androecium - absent
Gynoecium - same as disc floret

→ Fruit - cypsela
→ Seed - exalbuminous.

Genera Included under Asteraceae

- Aaronsohnia
- Abrotanella
- Acamptopappus
- Acanthocephalus
- Acanthocladium
- Acanthodesmos
- Acantholepis
- Acanthospermum
- Acanthostyles
- Achillea
- Achnophora
- Achnopogon
- Achyrachaena
- Achyrocline
- Achyropappus
- Achyrothalamus
- Acilepidopsis
- Acilepis
- Acmella
- Acomis
- Acourtia
- Acrisione
- Acritopappus
- Acroptilon
- Actinobole
- Actinoseris
- Actites
- Adelostigma
- Adenanthellum
- Adenocaulon
- Adenocritonia
- Adenoglossa
- Adenoon
- Adenopappus
- Adenophyllum
- Adenostemma
- Adenothamnus
- Aedesia
- Aegopordon
- Aequatorium
- Aetheolaena
- Aetheorhiza
- Ageratella
- Ageratina
- Ageratinastrum
- Ageratum
- Agoseris
- Agrianthus
- Ainsliaea
- Agania
- Aganiopsis
- Alatoseta
- Albertinia
- Alcantara
- Aliope
- Aldama
- Alepidocline
- Alfredia
- Aliella
- Allagopappus
- Allardia
- Allospermum
- Allopterigeron
- Almutaster
- Alomia
- Alomiella
- Alvordia
- Amauria
- Amberboa

Families of Dicotyledons

- → Amblyocarpum
- → Amblyolepis
- → Amblyopappus
- → Amboroa
- → Ambrosia
- → Ameghinoa
- → Amellus
- → Ammobium
- → Amolinia
- → Ampelaster
- → Amphiachyris
- → Amphiglossa
- → Amphipappus
- → Amphoricarpos
- → Anacantha
- → Anacyclus
- → Anaphalioides
- → Anaphalis
- → Anaxeton
- → Ancathia
- → Ancistrocarphus
- → Ancistrophora
- → Andryala
- → Angelphytum
- → Angianthus
- → Anisochaeta
- → Anisocoma
- → Anisopappus
- → Anisothrix
- → Antennaria
- → Anthemis
- → Antillia
- → Antiphiona
- → Antithrixia
- → Anura
- → Anvillea
- → Apalochlamys
- → Aphanactis
- → Aphanostephus
- → Aphyllocladus
- → Apodocephala
- → Apopyrus
- → Aposeris
- → Apostates
- → Arbelaezaster
- → Archibaccharis
- → Arctanthemum
- → Arctium
- → Arctogeron
- → Arctotheca
- → Arctotis
- → Argyranthemum
- → Argyroglottis
- → Argyrovernon
- → Argyroxiphium
- → Aristeguietia
- → Arnaldoa
- → Arnica
- → Arnicastrum
- → Arnoglossum
- → Arnoseris
- → Arrhenechthites
- → Arrojadocharis
- → Arrowsmithia
- → Artemisia
- → Artemisiopsis
- → Asaemia
- → Asanthus
- → Ascidiogyne
- → Aspilia
- → Asplundianthus
- → Aster
- → Asteridea

→ Asteriscus
→ Asteromoea
→ Asteropsis
→ Asterothamnus
→ Astranthium
→ Athanasia
→ Athrixia
→ Athroisma
→ Atractylis
→ Atractylodes
→ Atrichantha
→ Atrichoseris
→ Austrobrickellia
→ Austrochitonia
→ Austroeupatorium
→ Austrosynotis
→ Axiniphyllum
→ Ayapana
→ Ayapanopsis
→ Aylacophora
→ Aynia
→ Aztecaster
→ Baccharidopsis
→ Baccharis
→ Baccharoides

→ Badilloa
→ Baeriopsis
→ Bafutia
→ Bahia
→ Bahianthus
→ Baileya
→ Balduina
→ Balsamorhiza
→ Baltimora
→ Barkleyanthus
→ Barnadesia
→ Barroetea
→ Barrosoa
→ Bartlettia
→ Bartlettina
→ Basedowia
→ Bebbia
→ Bedfordia
→ Bejaranoa
→ Bellida
→ Bellis
→ Bellium
→ Belloa
→ Berardia
→ Berkheya
→ Berlandiera

→ Berroa
→ Berylsimpsonia
→ Bidens
→ Bigelowia
→ Bishopalea
→ Bishopanthus
→ Bishopiella
→ Bishovia
→ Blainvillea
→ Blakeanthus
→ Blakiella
→ Blanchetia
→ Blennosperma
→ Blennospora
→ Blepharipappus
→ Blepharispermum
→ Blepharizonia
→ Blumea
→ Blumeopsis
→ Boeberastrum
→ Boeberoides
→ Bolanosa
→ Bolocephalus
→ Boltonia
→ Bombycilaena

- → Borkonstia
- → Borrichia
- → Bothriocline
- → Brachanthemum
- → Brachionostylum
- → Brachyactis
- → Brachyclados
- → Brachyglottis
- → Brachylaena
- → Brachyscome
- → Brachythrix
- → Bracteantha
- → Brickellia
- → Brickelliastrum
- → Bryomorphe
- → Buphthalmum
- → Burkartia
- → Cabreriella
- → Cacalia
- → Cacaliopsis
- → Cacosmia
- → Cadiscus
- → Caesulia
- → Calea
- → Calendula
- → Callicephalus
- → Callilepis
- → Callistephus
- → Calocephalus
- → Calomeria
- → Calostephane
- → Calotesta
- → Calotis
- → Calycadenia
- → Calycoseris
- → Calyptocarpus
- → Camchaya
- → Camporassouria
- → Camptacra
- → Campuloclinium
- → Canadanthus
- → Cancrinia
- → Cancriniella
- → Cardopatium
- → Carduncellus
- → Carduus
- → Carlina
- → Carminatia
- → Carpesium
- → Carphephorus
- → Carphochaete
- → Carramiboa
- → Carterothamnus
- → Carthamus
- → Cassinia
- → Castanedia
- → Castalanthemu
- → Catanixis
- → Catananche
- → Catatia
- → Cavalcantia
- → Cavea
- → Celmisia
- → Centaurea
- → Centaurodendron
- → Centauropsis
- → Centaurothamnus
- → Centipeda
- → Centratherum
- → Cephalipterum
- → Cephalopappus
- → Cephalorrhynchu
- → Cephalosorus
- → Ceratogyne
- → Ceruana
- → Chacoa

→ Chaenactis
→ Chaetadelpha
→ Chaetanthera
→ Chaetopappa
→ Chaetoseris
→ Chamaechaenactis
→ Chamaegeron
→ Chamaemelum
→ Chamaepus
→ Chaptalia
→ Chardinia
→ Cheirolophus
→ Chersodoma
→ Chevreulia
→ Chiliadenus
→ Chiliocephalum
→ Chiliophyllum
→ Chiliotrichiopsis
→ Chiliotrichum
→ Chimantaea
→ Chionolaena
→ Chionopappus
→ Chlamydophora
→ Chloracantha
→ Chondrilla
→ Chondropyxis
→ Chresta
→ Chromolaena
→ Chromolepis
→ Chronopappus
→ Chrysactinia
→ Chrysactinium
→ Chrysanthellum
→ Chrysanthemoides
→ Chrysanthemum
→ Chrysanthoglossum
→ Chrysocephalum
→ Chrysocoma
→ Chrysogonum
→ Chrysolaena
→ Chrysoma
→ Chrysophthalmum
→ Chrysopsis
→ Chrysothamnus
→ Chthonocephalus
→ Chucoa
→ Chuquiraga
→ Cicerbita
→ Ciceronia
→ Cichorium
→ Cineraria
→ Cirsium
→ Cissampelopsis
→ Cladanthus
→ Cladochaeta
→ Clappia
→ Clibadium
→ Cnicothamnus
→ Cnicus
→ Coespeletia
→ Coleocoma
→ Coleostephus
→ Colobanthera
→ Columbiadoria
→ Comaclinium
→ Comborhiza
→ Commidendrum
→ Complaya
→ Condylidium
→ Condylopodium
→ Conoclinopsis
→ Conoclinium
→ Conyza
→ Coreocarpus
→ Coreopsis

- Corethamnium
- Correllia
- Corymbium
- Cosmos
- Cotula
- Coulterella
- Cousinia
- Cousiniopsis
- Craspedia
- Crassocephalum
- Cratystylis
- Cremanthus
- Croptilon
- Crossostephium
- Crossothamnus
- Crupina
- Cuatrecasanthus
- Cuatrecasasiella
- Cuchumatanea
- Cullumia
- Cuspidia
- Cyanthillium
- Cyathocline
- Cyathomone
- Cyclolepis
- Cylindrocline
- Cymbolaena
- Cymbonotus
- Cymbopappus
- Cynara
- Cyrtocymura
- Dacryotrichia
- Dahlia
- Damnamenia
- Damnxanthodium
- Dasycondylus
- Dasyphyllum
- Daveaua
- Decachaeta
- Decastylocarpus
- Decazesia
- Delairea
- Delamerea
- Delilia
- Dendranthema
- Dendrocacalia
- Dendrophorbium
- Dendrosenecio
- Dendroseris
- Denekia
- Desmanthodium
- Deuildemania
- Diacranthera
- Dianthoseris
- Diaphractanthus
- Diaspananthus
- Dicercoclados
- Dichaetophora
- Dichrocephala
- Dichromochlamys
- Dicoma
- Dicoria
- Dicranocarpus
- Didelta
- Dieletaia
- Digitacalia
- Dimeresia
- Dimerostemma
- Dimorphocoma
- Dimorphotheca
- Dinoseris
- Diodontium
- Diplazoptilon
- Diplostephium

- → Dipterocome
- → Dipterocypsela
- → Disparago
- → Dissothrix
- → Distephanus
- → Disynaphia
- → Dithyrostegia
- → Dittrichia
- → Doellingeria
- → Dolichoglottis
- → Dolichorrhiza
- → Dolichothrix
- → Doloniaea
- → Doniophyton
- → Dorobaea
- → Doronicum
- → Dracopis
- → Dresslerothamnus
- → Dubautia
- → Dubyaea
- → Dugesia
- → Duhaldea
- → Duidaea
- → Duseniella
- → Dymondia
- → Dyscritogyne
- → Dyscritothamnus
- → Dysodiopsis
- → Dyssodia
- → Eastwoodia
- → Eatonella
- → Echinacea
- → Echinocoryne
- → Echinops
- → Eclipta
- → Edmondia
- → Egletes
- → Eremocephala
- → Eitenia
- → Ekmania
- → Elachanthus
- → Elaphandra
- → Elephantopus
- → Eleutheranthera
- → Ellenbergia
- → Elytropappus
- → Embergeria
- → Emilia
- → Emiliella
- → Encelia
- → Enceliopsis
- → Endocellion
- → Endopappus
- → Engelmannia
- → Engleria
- → Enydra
- → Epaltes
- → Epilasia
- → Episcothamnus
- → Epitriche
- → Erato
- → Erechtites
- → Eremanthus
- → Eremosis
- → Eremothamnus
- → Eriachaenium
- → Ericameria
- → Ericentrodea
- → Erigeron
- → Eriocephalus
- → Eriochlamys
- → Eriophyllum

- Eriothrix
- Erlangea
- Erodiophyllum
- Erymophyllum
- Eryngiophyllum
- Erythradenia
- Erythrocephalum
- Espejoa
- Espeletia
- Espeletiopsis
- Ethulia
- Eucephalus
- Euchiton
- Eumorphia
- Eupatoriastrum
- Eupatorina
- Eupatoriopsis
- Eupatorium
- Euphrosyne
- Eurybiopsis
- Euryblochus
- Euryops
- Eutetras
- Euthamia
- Evacidium
- Ewartia
- Ewartiothamnus
- Exomiocarpon
- Facelis
- Farfugium
- Faujasia
- Faxonia
- Feddea
- Feldstonia
- Felicia
- Femeniasia
- Fenixia
- Ferreyranthus
- Ferreyrella
- Filago
- Filifolium
- Fitchia
- Fitzwillia
- Flaveria
- Fleischmannia
- Fleischmanniopsis
- Florestina
- Floscaldasia
- Flosmutisia
- Flourensia
- Flyriella
- Formania
- Foveolina
- Freya
- Fulcaldea
- Gaillardia
- Galactites
- Galeana
- Galeomra
- Galinsoga
- Gamochaeta
- Gamochaetopsis
- Garberia
- Garcibarrigoa
- Garcilassa
- Gardnerina
- Garhadiolus
- Garuleum
- Gazania
- Geigeria
- Geissolepis
- Geraea

- Gerbera
- Gerropogon
- Gibbaria
- Gilberta
- Gilruthia
- Gladiopappus
- Glaziovianthus
- Glossarion
- Glossocardia
- Glossopappus
- Glyptopleura
- Gnaphaliothamnus
- Gnaphalium
- Gnephosis
- Gochnatia
- Goldmanella
- Gongrostylus
- Gongylolepis
- Goniocaulon
- Gonospermum
- Gorceixia
- Gorteria
- Gossweilera
- Goyazianthus
- Grangea
- Grangeopsis
- Graphistyles
- Gratwickia
- Grauanthus
- Grazielia
- Greenmaniella
- Grindelia
- Grisebachianthus
- Grosvenoria
- Guardiola
- Guayania
- Guevaria
- Guizotia
- Gundelia
- Gundlachia
- Gutierrezia
- Gymnanthemum
- Gymnarrhena
- Gymnocondylus
- Gymnocoronis
- Gymnodiscus
- Gymnolaena
- Gymnopentzia
- Gymnosperma
- Gymnostephium
- Gynoxys
- Gynura
- Gypothamnium
- Gyptidium
- Gyptis
- Gysodoma
- Haastia
- Haeckeria
- Haegiela
- Handelia
- Haplocarpha
- Haploesthes
- Haplopappus
- Haplostephium
- Harleya
- Harnackia
- Hartwrightia
- Hasteola
- Hatschbachia
- Hazardia
- Hebeclinium

Families of Dicotyledons

- → Hecastocleis
- → Hedypnois
- → Helenium
- → Helianthella
- → Helianthus
- → Helichrysopsis
- → Helichrysum
- → Heliocauta
- → Heliomeris
- → Heliopsis
- → Helipterum
- → Helminthotheca
- → Hologyne
- → Hemisteptia
- → Hemizonia
- → Henricksonia
- → Heptanthus
- → Herderia
- → Herodotia
- → Herrickia
- → Hesperevax
- → Hesperodoria
- → Hesperomannia
- → Heteracia
- → Heteranthemis
- → Heterocoma
- → Heterocondylus
- → Heterocypsela
- → Heteroderis
- → Heterolepis
- → Heteromera
- → Heteromma
- → Heteropappus
- → Heteroplexis
- → Heterorhachis
- → Heterosperma
- → Heterothalamus
- → Heterotheca
- → Hidalgoa
- → Hieracium
- → Hilliardia
- → Hinterhubera
- → Hippia
- → Hippolytia
- → Hispidella
- → Hispidella
- → Hochstetteria
- → Hoehnephytum
- → Hoffmanniella
- → Hofmeisteria
- → Holocarpha
- → Holocheilus
- → Hololeion
- → Hololepis
- → Holozonia
- → Homognaphalium
- → Homogyne
- → Hoplophyllum
- → Huarpea
- → Hubertia
- → Hughesia
- → Hulsea
- → Humeocline
- → Hyalis
- → Hyalochaete
- → Hyalochlamys
- → Hyaloseris
- → Hyalosperma
- → Hybridella
- → Hydroidea
- → Hydropectis
- → Hymenocephalus

- Hymenoclea
- Hymenolepis
- Hymenonema
- Hymenopappus
- Hymenostemma
- Hymenothrix
- Hymenoxys
- Hyoseris
- Hypacanthium
- Hypericophyllum
- Hypochaeris
- Hysterionica
- Hystrichophora
- Ichthyothera
- Ideothamnus
- Ifloga
- Ighermia
- Iltisia
- Imeria
- Inezia
- Inula
- Inulanthera
- Inulopsis
- Iocenes
- Iodocephalus
- Iogeton
- Ionactis
- Iostephane
- Iotasperma
- Iphiona
- Iphionopsis
- Iranecio
- Irwinia
- Ischnea
- Ismelia
- Isocarpha
- Isocoma
- Isoetopsis
- Isostigma
- Iva
- Ixeridium
- Ixeris
- Ixeochlamys
- Ixiolaena
- Ixodia
- Jacmaia
- Jaegeria
- Jalcophila
- Jaliscoa
- Jamesianthus
- Jaramilloa
- Jasonia
- Jaumea
- Jefea
- Jeffreya
- Jessea
- Joseanthus
- Jungea
- Jurinea
- Jurinella
- Kalimeris
- Karelinia
- Karvandarina
- Kaschgaria
- Kaunia
- Keysseria
- Kinghamia
- Kingianthus
- Kippistia
- Kinkianella
- Kleinia
- Koanophyllon
- Koehneola
- Koelpinia
- Krigia

- Kyrsteniopsis
- Lachanodes
- Lachnophyllum
- Lachnorhiza
- Lachnospermum
- Lactacella
- Lactuca
- Lactucella
- Lactucosonchus
- Laennecia
- Laestadia
- Lagascea
- Lagedium
- Lagenithrix
- Lagenophora
- Laggera
- Lagophylla
- Lamprachaenium
- Lamprocephalus
- Lamyropappus
- Lamyropsis
- Langebergia
- Lantanopsis
- Lapsana
- Lapsanastrum
- Lasianthaea
- Lasiocephalus
- Lasiolaena
- Lasiopogon
- Lasiospermum
- Lasthenia
- Launaea
- Lawrencella
- Layia
- Lecocarpus
- Leibnitzia
- Leiboldia
- Lembertia
- Lemoorea
- Leontodon
- Leontopodium
- Lepidaploa
- Lepidesmia
- Lepidolopha
- Lepidolopsis
- Lepidonia
- Lepidophorum
- Lepidophyllum
- Lepidospartum
- Lepidostephium
- Leptinella
- Leptocarpha
- Leptoclinium
- Leptorhynchos
- Leptostelma
- Lescaillea
- Lessingia
- Lessingianthus
- Leucactinia
- Leucanthemella
- Leucanthemopsis
- Leucanthemum
- Leucheria
- Leucoblepharis
- Leucocyclus
- Leucogenes
- Leucomeris
- Leucophyta
- Leucoptera
- Leunisia
- Leuzea
- Leysera
- Liabellum
- Liabum
- Liatris

- Libanothamnus
- Ledbeckia
- Lifago
- Ligularia
- Limbarda
- Lindheimera
- Lopochaeta
- Litogyne
- Litothamnus
- Litrisa
- Llerasia
- Logfia
- Lomatozona
- Lonas
- Lopholaena
- Lophopappus
- Lordhowea
- Lorentzeanthus
- Loricaria
- Lourteigia
- Loxothysanus
- Lucilia
- Luciliocline
- Lugoa
- Luina
- Lulia
- Lundellianthus
- Lycapsus
- Lychnophora
- Lycoseris
- Lygodesmia
- Machaeranthera
- Macowania
- Macrachaenium
- Macraea
- Macroclinidium
- Macropodina
- Macvaughiella
- Madagaster
- Madia
- Mairia
- Malacothrix
- Mallotopus
- Malmeanthus
- Malperia
- Mantisalca
- Marasmodes
- Marshallia
- Marshalljohnstonia
- Mastricorenia
- Matricaria
- Mattfeldanthus
- Mattfeldia
- Matudina
- Mauranthemum
- Mausolea
- Mecomischus
- Megalodonta
- Melampodium
- Melanodendron
- Melanthera
- Merrittia
- Metalasia
- Metastevia
- Mexerion
- Mexianthus
- Micractis
- Microcephala
- Microglossa
- Microgynella
- Microliabum
- Micropsis
- Micropus
- Microseris
- Mikania
- Microspermum
- Mikaniopsis
- Milleria

- Melletia
- Minuria
- Mexicacalia
- Meyamayomena
- Mniodes
- Monactis
- Monarrhenus
- Monogereion
- Monolopia
- Monoptelon
- Montanoa
- Monticalia
- Moonia
- Moquinia
- Morithamnus
- Moscharia
- Msuata
- Mulgedium
- Munnozia
- Munzothamnus
- Muschleria
- Mutisia
- Mycelis
- Myopordon
- Myriactis
- Myriocephalus
- Myripnois
- Myxopappus
- Nabalus
- Nananthea
- Nannoglottis
- Nanothamnus
- Nardophyllum
- Narvalina
- Nassauvia
- Nauplius
- Neblenaea
- Neja
- Nelsonianthus
- Nemosenecio
- Neocabreria
- Neocuatrecasia
- Neohintonia
- Neojeffreya
- Neomirandea
- Neomolina
- Neopallasia
- Neotysonia
- Nesomia
- Nestlera
- Neurolaena
- Neurolakis
- Nicolasia
- Nicolletia
- Nidorella
- Niketenia
- Nipponanthemu
- Nivellea
- Nolletia
- Nothobaccharis
- Nothocalais
- Noticastrum
- Notobasis
- Notoseris
- Nouelia
- Novenia
- Oaxacania
- Oblivia
- Ochrocephala
- Oclemena
- Odixia
- Odontocline
- Oedera
- Oiospermum
- Oldenburgia
- Olearia

- Olgaea
- Oligactis
- Oliganthes
- Oligocarpus
- Oligochaeta
- Oligoneuron
- Oligothrix
- Olivaea
- Omalotheca
- Omphalopappus
- Oncosiphon
- Ondetia
- Onopordum
- Onoseris
- Oonopsis
- Oparanthus
- Ophryosporus
- Opisthopappus
- Oreochrysum
- Oreoleysera
- Oreostemma
- Oritrophium
- Orochaenactis
- Osbertia
- Osmadenia
- Osmiopsis
- Osmitopsis
- Osteospermum
- Otanthus
- Oteiza
- Othonna
- Otopappus
- Otospermum
- Outreya
- Oxycarpha
- Oxylaena
- Oxylobus
- Oxypappus
- Oxyphyllum
- Oyedaea
- Ozothamnus
- Pachylaena
- Pachystegia
- Pachythamnus
- Pacifigeron
- Packera
- Pacourina
- Palafoxia
- Paleaepappus
- Pamphalea
- Pappobolus
- Pappochroma
- Paracalia
- Parachionolaena
- Paragynoxys
- Paralychnophora
- Paranephelius
- Parantennaria
- Parapiqueria
- Paraprenanthe
- Parasenecio
- Parastrephia
- Parthenice
- Parthenium
- Pasaccardoa
- Pechuelloeschea
- Pectis
- Pegolettia
- Pelucha
- Pentacalia
- Pentachaeta
- Pentanema
- Pentatrichia

- Pentzia
- Perdicium
- Perezia
- Pericallis
- Pericome
- Peripleura
- Perystyle
- Perralderia
- Pertya
- Perymeniopsis
- Perymenium
- Petalacte
- Petasites
- Peteravenia
- Petradoria
- Petrobium
- Peucephyllum
- Phacellothrix
- Phaenocoma
- Phaeostigma
- Phagnalon
- Phalacrachena
- Phalacraea
- Phalacrocarpum
- Phalacroseris
- Phanomon
- Phicradeniopsis
- Picris
- Picrosia
- Picrothamnus
- Pilosella
- Pilostemon
- Pinaropappus
- Pingraea
- Pinillosia
- Piora
- Pippenalia
- Piptocarpha
- Piptocoma
- Piptolepis
- Piptothrix
- Piqueria
- Piqueriella
- Piqueriopsis
- Pithecoseris
- Pithocarpa
- Pittocaulon
- Pityopsis
- Pladaroxylon
- Plagiobasis
- Plagiocheilus
- Plagiolophus
- Plagius
- Planaltoa
- Planea
- Plateilema
- Platycarpa
- Platypodanthera
- Platyschkuhria
- Plazia
- Plecostachys
- Plectocephalus
- Pleiotaxis
- Pleocarphus
- Pleurocarpaea
- Pleurocoronis
- Pleurophyllum
- Pluchea
- Podachaenium
- Podanthus
- Podocoma
- Podolepis

- Podotheca
- Poecilolepis
- Pogonolepis
- Poparkonia
- Pollalesta
- Polyachyrus
- Polyanthina
- Polyarrhena
- Polycalymma
- Polychrysum
- Polymnia
- Polytaxis
- Porophyllum
- Porphyrostemma
- Praxeliopsis
- Praxelis
- Prenanthella
- Prenanthes
- Printzia
- Prionopsis
- Prolobus
- Prolongoa
- Proteopsis
- Proustia
- Psacaliopsis
- Psacalium
- Psathyrotes
- Psathyrotopsis
- Psednotrichia
- Pseudelephantopus
- Pseudobahia
- Pseudoblepharispermum
- Pseudobrickellia
- Pseudocadiscus
- Pseudoclappia
- Pseudoconyza
- Pseudognaphalium
- Pseudogynoxys
- Pseudohandelia
- Pseudojacobaea
- Pseudokyrsteniopsis
- Pseudoligandra
- Pseudonoseris
- Pseudostifftia
- Psiadia
- Psiadiella
- Psilactis
- Psilocarphus
- Psilostrophe
- Psychrogeton
- Psychrophyton
- Pterachaenia
- Pterocaulon
- Pterocypsela
- Pteronia
- Pterothrix
- Pterygopappus
- Ptilostemon
- Pulicaria
- Pycnocephalus
- Pyrrhopappus
- Pyrrocoma
- Pytinicarpa
- Quelchia
- Quinetia
- Quinqueremu
- Radlkoferotoma
- Rafinesquia
- Raillardella
- Raillardeopsis

- Rainiera
- Raoulia
- Raoulinopsis
- Rastrophyllum
- Ratibida
- Raulinoreitzia
- Rayjacksonia
- Reichardia
- Relhania
- Remya
- Rennera
- Rensonia
- Revealia
- Rhagodiolus
- Rhamphogyne
- Rhanteriopsis
- Rhanterium
- Rhetinolepis
- Rhodanthe
- Rhodanthemum
- Rhynchopsidium
- Rhynchospermum
- Rhysolepis
- Richteria
- Riencourtia
- Rigiopappus
- Robinsonecio
- Robinsonia
- Rochonia
- Rojasianthe
- Rolandra
- Rosenia
- Rothmaleria
- Rudbeckia
- Rugelia
- Ruilopezia
- Rumfordia
- Russowia
- Rutidosis
- Sabazia
- Sachsia
- Salmea
- Santolina
- Santosia
- Sanvitalia
- Sarcanthemum
- Sartorina
- Sartwellia
- Saussurea
- Scalesia
- Scaevola
- Schenya
- Schischkinia
- Schistocarpha
- Schistostephium
- Schizogyne
- Schizoptera
- Schizotrichia
- Schkuhria
- Schlechtendalia
- Schmalhausenia
- Schoenia
- Sciadocephala
- Sclerocarpus
- Sclerolepis
- Sclerorhachis
- Sclerostephane
- Scolymus
- Scorzonera
- Scrobicaria
- Selloa
- Senecio
- Sericocarpus
- Seriphidium

- → Serratula
- → Shafera
- → Sheareria
- → Shinnersia
- → Shinnerxoseris
- → Spapaea
- → Siebera
- → Sigesbeckia
- → Silloxerus
- → Silphium
- → Silybum
- → Simsia
- → Sinacalia
- → Sinclairia
- → Sinoleontopodium
- → Sinosenecio
- → Sippolesia
- → Smallanthus
- → Soaresia
- → Solanecio
- → Solenogyne
- → Solidago
- → Soliva
- → Sommerfeltia
- → Sonchus
- → Sondottia
- → Soroseris
- → Spaniopappus
- → Sparganophorus
- → Sphaeranthus
- → Sphaereupatorium
- → Sphaeromeria
- → Sphagneticola
- → Spilanthes
- → Spiracantha
- → Spiroseris
- → Squamopappus
- → Stachycephalum
- → Staehelina
- → Standleyanthus
- → Staurochlamys
- → Stebbinsoseris
- → Steiractinia
- → Steirodiscus
- → Stemmacantha
- → Stenachaenium
- → Stenocephalum
- → Stenocline
- → Stenopadus
- → Stenophalium
- → Stenops
- → Stenoseris
- → Stenotus
- → Stephanochilus
- → Stephanodoria
- → Stephanomeria
- → Steptorhamphus
- → Stevia
- → Steviopsis
- → Steyermarkina
- → Stifftia
- → Stilpnogyne
- → Stilpnolepis
- → Stilpnopappus
- → Stoebe
- → Stokesia
- → Stomatanthes
- → Stomatochaeta
- → Stramentopappus
- → Streptoglossa
- → Strotheria
- → Stuartina
- → Stuckertiella

Families of Dicotyledons

- Stuessya
- Stylocline
- Stylotrichium
- Sventenia
- Symphyllocarpus
- Symphyopappus
- Symphyotrichum
- Syncalathium
- Syncarpha
- Syncephalum
- Syncretocarpus
- Synedrella
- Synedrellopsis
- Syneilesis
- Synotis
- Syntrichopappus
- Synurus
- Syreitschikovia
- Taeckholmia
- Tagetes
- Takeikadzuchia
- Takhtajaniantha
- Talamancalia
- Tamananthus

- Tamania
- Tamaulepa
- Tanacetopsis
- Tanacetum
- Taplinia
- Taraxacum
- Tarchonanthus
- Tehuana
- Teixeiranthus
- Telanthophora
- Telekia
- Telmatophila
- Tenrhynea
- Tephroseris
- Tessaria
- Tetrachyron
- Tetradymia
- Tetragonotheca
- Tetranocolopium
- Tetraneuris
- Tetranthus
- Tetraperone
- Thaminophyllum
- Thamnoseris

- Thelesperma
- Thespidium
- Thespis
- Thevenotia
- Thiseltonia
- Thurovia
- Thymophylla
- Thymopsis
- Tiarocarpus
- Tietkensia
- Tithonia
- Tolbonia
- Tolpis
- Tomentaurum
- Tonestus
- Tourneuxia
- Townsendia
- Tracyina
- Tragopogon
- Traversia
- Trechanthemis
- Trichanthodium
- Trichocline
- Trichocoronis
- Trichocoryne

- → Trichogonia
- → Trichogoniopsis
- → Trichogyne
- → Trecholepis
- → Trichoptileum
- → Trichospira
- → Tridactylina
- → Tridax
- → Trigonospermum
- → Tritisa
- → Trimorpha
- → Triuncinia
- → Tripleurospermum
- → Triplocephalum
- → Tripteris
- → Triptilion
- → Triptilodiscus
- → Trixis
- → Troglophyton
- → Tuberostylis
- → Tugarinovia
- → Turaniphytum
- → Tussilago
- → Tuxtla
- → Tyleropappus
- → Tyrimnus
- → Uechtritzia
- → Ugamia
- → Uleophytum
- → Uncia
- → Urbananthus
- → Urbinella
- → Urmenetea
- → Urolepis
- → Uropappus
- → Urospermum
- → Ursinia
- → Vanclevea
- → Varilla
- → Varthemia
- → Vellereophyton
- → Venegasia
- → Verbesina
- → Vernonia
- → Vernoniopsis
- → Viereckia
- → Vieria
- → Vigethia
- → Viguiera
- → Villanova
- → Vilobia
- → Vittadinia
- → Vittetia
- → Volutaria
- → Waitzia
- → Wamalchitamia
- → Warionia
- → Wedelia
- → Welwitschiella
- → Wendelboa
- → Werneria
- → Westoniella
- → Whitneya
- → Wilkesia
- → Willemetia
- → Wollastonia
- → Wulffia
- → Wunderlichia
- → Wyethia
- → Xanthisma
- → Xanthium
- → Xanthocephalum

- Xanthopappus
- Xeranthemum
- Xerolekia
- Xylanthemum
- Xylorhiza
- Xylothamia
- Yermo
- Youngia
- Zaluzania
- Zandera
- Zexmenia
- Zinnia
- Zoegea
- Zyrphelis
- Zyzyxia

Sub class: Gamopetalae

Series: Heteromerae

Order: Ericales

Families:

Ericaceae

Vaccineae

Monotropeae

Epacrideae

Diapensiaceae

Lennoaceae

Chapter-92: Ericaceae

Classification (Bentham and Hooker)

- Phanerogams
- Dicotyledons
- Gamopetalae
- Heteromerae
- Ericales
- Ericaceae

General characters

→ Mostly shrubs rarely herbs and trees
→ Leaves - alternate, evergreen
→ Plants hermaphrodite
→ Flowers - regular
→ Sepals 5 united at the base and petals 5 united. rarely 4 or more or less. often bell shaped; white to pink or red in colour.
→ Stamens - twice as petals.
→ Ovary - superior or inferior; usually 5 or 4 united carpels with partition walls, forming an equal number of chambers.
→ Fruit - capsule, berry, drupe.

Genera included under Ericaceae

- Enkianthus
- Chimaphila
- Moneses
- Orthilia
- Pyrola
- Allotropa
- Cheilotheca
- Hemitomes
- Monotropa
- Monotropastrum
- Monotropsis
- Pityopus
- Pleuricospora
- Pterospora
- Sarcodes
- Arbutus
- Arctostaphylos
- Comarostaphylis
- Ornithostaphylos
- Xylococcus
- Cassiope
- Bryanthus
- Ledothamnus
- Ceratiola
- Corema
- Empetrum
- Calluna
- Daboecia
- Erica
- Bejaria
- Elliottia
- Epigaea
- Kalmia
- Kalmiopsis
- Phyllodoce
- Rhodothamnus
- Rhododendron
- Therorhodion
- Harrimanella
- Archeria
- Andersonia
- Cosmelia
- Sprengelia
- Budawangia
- Epacris
- Lysinema
- Rupicola
- Woollsia
- Dielsiodoxa
- Needhamiella
- Oligarrhena
- Lebetanthus
- Prionotes
- Dracophyllum
- Richea
- Sphenotoma
- Acrothamnus
- Acrotriche
- Agiortia
- Androstoma
- Astroloma
- Brachyloma
- Coleanthera
- Conostephium
- Croninia
- Cyathodes
- Cyathopsis
- Decatoca

Families of Dicotyledons

- → Leptecophylla
- → Leucopogon
- → Lissanthe
- → Melichrus
- → Monotoca
- → Montitega
- → Pentachondra
- → Planocarpa
- → Styphelia
- → Trochocarpa
- → Andromeda
- → Zenobia
- → Chamaedaphne
- → Oepplycosia
- → Eubotrys
- → Gaultheria
- → Leucothoe
- → Pernettya
- → Tepuia
- → Agarista
- → Craibiodendron
- → Lyonia
- → Pieris
- → Oxydendrum
- → Agapetes
- → Anthopteropsis
- → Anthopterus
- → Cavendishia
- → Ceratostema
- → Costera
- → Demosthenesia
- → Didonica
- → Dimorphanthera
- → Diogenesia
- → Disterigma
- → Gaylussacia
- → Gonocalyx
- → Lateropora
- → Macleania
- → Mycerinus
- → Notopora
- → Oreanthes
- → Orthaea
- → Paphia
- → Pellegrinia
- → Plutarchia
- → Polyclita
- → Psammisia
- → Rusbya
- → Satyria
- → Semiramisia
- → Siphonandra
- → Sphyrospermum
- → Symphysia
- → Themistoclesia
- → Thibaudia
- → Utleya
- → Vaccinium

Chapter - 93
Vacinieae

Classification (Bentham and Hooker)

Phanerogams
Dicotyledons
Gamopetalae
Heteromerae
Ericales
Vaccinieae

General characters

→ Shrubs or trees, usually evergreen
→ Leaves - Simple, alternate, opposite or whorled; exstipulate
→ Flowers - regular, hermaphrodite
→ Calyx - superior or inferior; limb 4 or 5 lobed
→ Corolla - campanulate or urceolate; lobes imbricate
→ Stamens 4-10, hypo or epigynous; anthers - 2 celled; often furnished with an awn like appendage
→ Disk - annular, lobed or glandular
→ Fruit - berry or capsule

Genera : Now included under Ericaceae

Chapter - 94
Monotropeae

Classification (Bentham and Hooker)

Phanerogams
Dicotyledons
Gamopetalae
Heteromerae
Ericales
Monotropeae

General characters

→ Subshrubs, shrubs or Trees
→ Inflorescence - Raceme, spike, rarely solitary
→ Shoots are achlorophyllous, hence plants are non-photosynthetic and have striking and distinctive appearance, with coloration ranging from pure white to bright yellow or red.
→ Corolla - bell or cup shaped.
→ Petals - may or may not be fused.
→ Poricidal anthers
→ Pollen grains are released as a monad.
→ Fruits - dry loculicidal dehiscent, capsule or berry.

Genera included under Monotropeae

- Allotropa
- Cheilotheca
- Hemitomes
- Hypopitys
- Monotropa
- Monotropastrum
- Monotropsis
- Pityopus
- Pleuricospora

Chapter - 95
Epacridaceae

Classification (Bentham and Hooker)

- Phanerogams
- Dicotyledons
- Gamopetalae
- Heteromerae
- Ericales
- Epacridaceae

General characters

→ Small trees or shrubs
→ Leaves - persistent, alternate, spiral, flat or coiled, herbaceous or leathery, petiolate to sessile; sheathing or not; leaf sheath when present with free margins; leaves - simple; exstipulate
→ Plants hermaphrodite. Pollination entomophilous or ornithophilous
→ Flowers solitary or aggregated in inflorescence
→ Flowers - bracteate, bracteolate, regular, tetra or pentamerous, tetracyclic. Hypogynous disk usually present
→ Perianth with distinct calyx and corolla;

Calyx consists of 4 or 5 sepals; 1 whorled, Poly sepalous, regular, Persistent, imbricate.

→ Corolla – 4 or 5 petals, 1 whorled; Poly or gamo petalous. imbricate or valvate or contorted; regular; green or white or red / pink / purple / blue / yellow.

→ Androecium 4-5 stamens or rarely 2; epipetalous; oppositisepalous; anthers basifixed or adnate; non-versatile, dehiscing via longitudinal slits.

→ Gynoecium 2-10 carpelled., syncarpous; ovary 1-10 celled. Gynoecium stylate; styles-1, stigma-1

→ Fruit – fleshy or non-fleshy, dehiscent or indehiscent.; capsule or drupe

→ Seeds – endospermic (oily); Embryo straight

Genera Included under Epacrideae

- Acrothamnus
- Acrotriche
- Andersonia
- Androstoma
- Archeria
- Astroloma
- Brachyloma
- Choristemon
- Coleanthera
- Conostephium
- Cosmelia
- Cyathodes
- Cyathopsis
- Decatoca
- Dracophyllum
- Epacris
- Lebetanthus
- Leptecophylla
- Leucopogon
- Lissanthe
- Lysinema

- → Melichrus
- → Monotoca
- → Needhamiella
- → Oligarrhena
- → Pentachondra
- → Planocarpa
- → Prionotes
- → Richea
- → Rupicola
- → Sphenotoma
- → Sprengelia
- → Styphelia
- → Trochocarpa
- → Woolsia

Chapter - 96
Diapensiaceae
(Bentham and Hooker)

Classification
Phanerogams
Dicotyledons
Gamopetalae
Heteromerae
Ericales
Diapensiaceae

General characters

→ Small shrubs or herbs
→ Leaves - alternate, spiral, petiolate to sessile; non-sheathing, simple; Lamina entire; Margin - entire or serrate or dentate; exstipulate.
→ Plants hermaphrodite
→ Flowers solitary or in inflorescence - racemes.
→ Flowers - bracteolate, regular, pentamerous, tetra or pentacyclic
→ Perianth with distinct calyx and corolla.
→ Calyx consists of 5 sepals in 1 whorl; Poly or gamo sepalous; when gamosepalous 5 blunt-lobed, regular, persistent, imbricate.
→ Corolla consists of 5 petals in 1 whorl; Polypetalous

or gamopetalous; imbricate or contorted; regular; white or pink or purple; Petals deeply bifid or belobed or fringed or entire.

→ Androecium - 5 or 10 stamens; free of one another to coherent; usually isomerous with the perianth, rarely diplostemonous; Anthers - basifixed; dehiscing via longitudinal slits or transversely.

→ Gynoecium - Tricarpellary, Syncarpous, Superior; ovary 3 locular; Style - 1, attenuate from the ovary, apical; stigma - 1, 3 lobed. Axile placentation; ovules 5-50 per locule

→ Fruit - non-fleshy, dehiscent, a capsule (loculicidal)

→ Seeds endospermic; embryo - straight to curved

Genera included under Diapensiaceae

- → Berneuxia
- → Diapensia
- → Galax
- → Pyxidanthera
- → Schizocodon
- → Shortia

Chapter - 97

Lennoaceae

Classification (Bentham and Hooker)

Phanerogams
Dicotyledons
Gamopetalae
Heteromerae
Ericales
Lennoaceae

General characters

→ Herbs; achlorophyllous; Plants succulent; Parasitic.

→ Leaves - small, alternate, spiral, reduced to short scales (Membranous), simple; lamina - entire; exstipulate

→ Plants hermaphrodite

→ Inflorescence - Panicles or heads or spikes; with or without involucral bracts.

→ Flowers - bracteate, regular or somewhat irregular; slightly zygomorphic; 5-10 merous; Tetra or Pentacyclic.

→ Perianth with distinct calyx and corolla;

→ Calyx consists of 5-10 sepals; 1 whorled; gamo or polysepalous. Corolla consists of 5-10 petals, 1 whorled

gamopetalous; valvate or imbricate;
→ Androecium – 5-10 stamens; 1-2 whorled; Stamens inserted in the throat of the corolla tube; isomerous with the perianth; Anthers introrse; dehisce via longitudinal slits.
→ Gynoecium – 6 to 14 carpelled; syncarpous; superior; Ovary 6-14 locular. Locules secondarily divided by 'false septa'. Style-1, apical. Stigma-1 (capitate or lobed); Placentation axile; ovules 2 per locule.
→ Fruit – fleshy or non fleshy; dehiscent, capsule
→ Seeds endospermic (oily)

Genera included under Lennoaceae
→ Ammobroma
→ Lennoa
→ Pholisma

Sub class: Gamopetalae

Series: Heteromerae

Order: primulales

Families:

Plumbagineae

Primulaceae

Myrsineae

Chapter – 98
Plumbaginaceae

Classification (Bentham and Hooker)
Phanerogams
Dicotyledons
Gamopetalae
Heteromerae
Primulales
Plumbaginaceae

General characters

→ Herbs; Mostly Perennial; shrubs or leanas; self supporting or climbing; sometimes stem twiners.

→ Leaves – alternate, spiral, herbaceous or leathery; Petiolate to sessile; simple; epulvinate; lamina – dissected or entire; often acicular, or linear or oblong; stipulate or exstipulate.

→ Plants hermaphrodite.

→ Pollination – entomophilous.

→ Inflorescence – panicles or heads or racemes.

→ Flowers – bracteolate; small; regular; Pentamerous; tetracyclic.

→ Perianth with distinct calyx and corolla.

→ Calyx consists of 5 sepals in 1 whorl; gamo

sepalous; blunt lobed or toothed; regular, persistent; valvate or plicate
→ Corolla consists of 5 petals, 1 whorled; gamo or polypetalous; imbricate or contorted; regular; white/yellow/red/pink/purple/blue. fleshy or leathery; often persistent
→ Androecium - 5 stamens - 1 whorled, isomerous with perianth; Anthers - dorsifixed or basifixed; introrse, dehiscing via longitudinal slits.
→ Gynoecium - 5 carpellary, syncarpous, superior; Ovary 1 locular; Style 1 or 5, free to partly joined, apical; Stigma - dry type. Placentation basal. 1 ovule per cavity; anatropous.
→ Fruit - dehiscent, or indehiscent; capsule or nut.
→ Seeds endospermic or non-endospermic; winged.

Genera included under Plumbagineae

- Acantholimon
- Aegialitis
- Armeria
- Bamiana
- Bucinicza
- Cephalorhizum
- Ceratostigma
- Chaetolimon
- Dictyolimon
- Dyerophytum
- Eremolimon
- Ghasnianthus
- Goniolimon
- Ikonnikovia
- Limoniastrum
- Limoniopsis
- Limonium
- Meullerolimon
- Neogontscharovia
- Plumbagella
- Plumbago
- Popoviolimon
- Psylliostachys
- Vassilczenkoa

Chapter - 99
Primulaceae

Classification (Bentham and Hooker)

Phanerogams
Dicotyledons
Gamopetalae
Heteromerae
Primulales
Primulaceae

General characters

→ Herbs; Mostly perennial; often rhizomatous or tuberous. Hydrophytic or halophytic or xerophytic.

→ Leaves - alternate or opposite or whorled; when alternate, spiral; petiolate to sessile; non sheating; rarely gland dotted (Anagallis). Lamina mostly entire; exstipulate; Lamina Margin crenate to dentate rarely entire

→ Plants hermaphrodite. Pollination entomophilous.

→ Flowers solitary or aggregated in inflorescence - heads or umbels or panicles; often scapiflorus.

→ Flowers - ebracteolate, Medium sized or regular, 3-9 merous, tetra or pentacyclic

- Perianth with distinct calyx and corolla (or) sepaline. Calyx consists of 3-9 sepals in 1 whorl; gamosepalous, regular, usually persistent; imbricate or contorted.
- Corolla 3-9 petals in 1 whorl; gamopetalous; imbricate or contorted; regular; green or white/ yellow/ red / purple/ blue.
- Androecium 3-10 stamens, 1 or 2 whorled; Stamens inserted near the base of the corolla tube. Anthers dehisce via pores or longitudinal slits.
- Gynoecium - 5 carpelled; isomerous with the perianth. Syncarpous, usually superior rarely partly inferior; ovary 1 locular; styles - 1 apical; stigma - 1, dry type; placentation - free central; 5 - many ovules per cavity.
- Fruit - non fleshy, dehiscent or rarely indehiscent, capsule
- Seeds - endospermic.

Genera included under Primulaceae
- Anagallis
- Androsace
- Ardisiandra
- Bryocarpum
- Cortusa
- Cyclamen
- Dionysia
- Dodecatheon
- Glaux
- Hottonia
- Kaufmannia
- Lysimachia
- Omphalogramma
- Pelletiera
- Pomatosace
- Primula
- Samolus
- Soldanella
- Stimpsonia
- Trientalis

Chapter – 100
Myrsineae

Classification (Bentham and Hooker)

Phanerogams
Dicotyledons
Gamopetalae
Heteromerae
Primulales
Myrsineae

General characters

→ Trees and shrubs; few lianas; few sub-herbaceous. few climbing or mostly self supporting.

→ Leaves - alternate, spiral, petiolate, non-sheating; some aromatic; simple; lamina entire; margin entire; exstipulate. Domatia occurring in the family manifested as pockets or hair tufts.

→ Plants hermaphrodite or monoecious or dioecious

→ Female flowers often with staminodes.

→ Inflorescence - racemose; axillary, terminal;

→ Flowers - Mostly ebracteolate, rarely bracteolate. small, regular. tetra or pentamerous;

tetracyclic
- → Perianth with distinct calyx and corolla
- → Calyx consists of 3-6 sepals in 1 whorl, poly or gamosepalous (basally connate); regular; imbricate or contorted or valvate.
- → Corolla consists of 3-6 petals in 1 whorl; gamopetalous rarely polypetalous; imbricate / contorted / valvate
- → Androecium 3-6 stamens; free of one another or monadelphous; Anthers cohering (Amblyanthus) or free; dehisce via pores or longitudinal slits; introrse
- → Gynoecium 3-6 carpelled; syncarpous, superior or partly inferior; Ovary 1 locular; styles - 1, apical; stigma - 1. Placentation - basal or free central; ovules 3 - many per cavity; sunken in the placenta
- → Fruit - fleshy, indehiscent, berry or drupe
- → Seeds endospermic

Genera Included under Myrsineae

- Aegiceras
- Amblyanthopsis
- Amblyanthus
- Anagallis
- Antistrophe
- Ardisia
- Asterolinon
- Badula
- Conandrium
- Coris
- Ctenardisia
- Cybianthus
- Cyclamen
- Discocalyx
- Elingamita
- Embelia
- Emblemantha
- Fittingia
- Geissanthus
- Glaux
- Heberdenia
- Hymenandra
- Labisia
- Loheria
- Lysimachia
- Maesa
- Monoporus
- Myrsine
- Oncostemum
- Parathesis
- Pelletiera
- Pleiomeris
- Rapanea
- Sadiria
- Solonia
- Stylogyne
- Tapeinosperma
- Trientalis
- Tetrardisia
- Vegaea
- Wallenia

Sub class: Gamopetalae

Series: Heteromerae

Order: Ebenales

Families:

Sapotaceae

Ebenaceae

Styraceae

Chapter-101
Sapotaceae

Classification (Bentham and Hooker)

- Phanerogams
- Dicotyledons
- Gamopetalae
- Inferae
- Ebenales
- Sapotaceae

General Characters

→ Mostly trees and rarely shrubs
→ Milky latex is found in pith, cortical region and in leaves. Leaves, flowers and fruits are usually covered with unicellular 2-armed hairs.
→ Leaves - Simple, alternate, rarely sub opposite, petiolate, coriaceous, entire, exstipulate or very caducous.
→ Inflorescence - Flowers solitary or in cymose inflorescence. In some cases, inflorescence is found to be situated above the scars of fallen leaves or on old stems
→ Flowers - hermaphrodite, actinomorphic, bracteolate, bracteoles when found are small and caducous, complete and hypogynous.
→ The calyx consists of sepals which show considerable

Variation eg:- 2+2, 3+3, 4+4 or 5. Sepals are somewhat connate at the base (gamosepalous). Aestivation - imbricate

→ Corolla - the number of petals corresponds to the number of sepals; the petals alternate the sepals. Mostly arranged in single whorl, rarely in 2 whorls. Petals are united basally. Number of petals apparently increase because of the development of dorsal outgrowths from the base of each petal in Mimusops. These outgrowths resemble the petals. In other genera, each segment is found with a pair of lateral outgrowths.

→ Androecium - stamens are epipetalous (ie) situated on corolla tube. Generally arranged in 2 or 3 whorls each of 4-5 stamens. The inner whorls consists of fertile stamens and the stamens of remaining whorls are being reduced to staminodes. or sometimes altogether suppressed. Staminodes remain alternate to stamens, filaments usually short; anthers oblong-lanceolate, connective often produced, anthers 2 celled, mostly extrorse, dehisce longitudinally.

→ Gynoecium consists of as many carpels or double as many as the number of stamens in a whorl; ovary superior, tetra, penta or multichambered;

Placentation axile; one anatropous ovule in each loculus; style simple, stigma inconspicuous
- → Fruit is a berry — outer layer is brown and sometimes sclerenchymatous. Berry is sticky in nature which enable its distribution by birds
- → Seeds are endospermic or non-endospermic when the cotyledons are flat, when non-endospermic the cotyledons are large and fleshy, testa is hard and shiny.
- → Pollination is entomophilous.

Genera included under Sapotaceae

- Achras
- Argania
- Aubregrinia
- Aulandra
- Autranella
- Baillonella
- Brevica
- Bumelia
- Burckella
- Butyrospermum
- Calocarpum
- Capurodendron
- Chromolucuma
- Chrysophyllum
- Delpydora
- Diploknema
- Dipholis
- Diploon
- Eberhardtia
- Ecclinusa
- Elaeoluma
- Englerophytum
- Faucherea
- Gluema
- Inhambanella
- Isonandra
- Labourdonnaisia
- Labramia
- Lecomtedoxa
- Leptostyles
- Letestua
- Madhuca
- Manilkara
- Micropholis
- Mimusops
- Neohemsleya
- Neolemonniera
- Nesoluma
- Niemeyera
- Northia
- Omphalocarpum
- Palaquium
- Payena
- Pichonia
- Pouteria
- Pradosia
- Pycnandra
- Sarcaulus
- Sideroxylon
- Tieghemella
- Tridesmostemon
- Tsebona
- Vitellaria
- Vitellariopsis
- Xantolis

Chapter - 102
Ebenaceae

Classification (Bentham and Hooker)

- Phanerogams
- Dicotyledons
- Gamopetalae
- Inferae
- Ebenales
- Ebenaceae

General Characters

→ Trees or shrubs with hard wood in stems. Heartwood is often black, red or green; Milky sap is absent; secretary cells with tannin are found in the tissues of stem

→ Leaves - alternate, rarely opposite or whorled, simple, entire, coriaceous, exstipulate.

→ Inflorescence - cymose or solitary axillary.

→ Flowers - unisexual usually in dioecious plants, rarely flowers are bisexual and polygamous, actinomorphic, 3-7 merous, hypogynous, staminate flowers mostly more abundant than pistillate.

→ Calyx - sepals 3-7 usually 4-5, united in persistent lobed calyx (gamosepalous)

→ Corolla - petals 3-7, gamopetalous, united in an

urceolate, tubular, campanulate or semi-funnel-shaped corolla, lobes imbricate or contorted in bud.

→ Androecium – stamens epipetalous or hypogynous, 3-7 or 2-3 times as many as corolla, distinct or united in pairs, anthers dithecous (2 celled), introrse, dehisce longitudinally, staminodes & pistillode are present in female and male flowers respectively.

→ Gynoecium – carpels 2-16, usually 2-8, syncarpous, superior ovary, with locules as many as carpels, false septum frequently develops in ovary. Ovules 2-1, integuments 2, ovules pendulous and anatropous; styles and stigmas 2-8, styles distinct or basally connate, Axile placentation; multilocular superior ovary.

→ Fruit – A berry with one or many seeds.

→ Seeds with straight embryo and very hard, copious endosperms.

Secretory cells containing tannin like substance occur in the tissues of leaves and stems. Various kinds of hairs are present on the leaves.

Genera included under Ebenaceae
→ Diospyros
→ Euclea

Chapter-103
Styraceae

Classification (Bentham and Hooker)

- Phanerogams
- Dicotyledons
- Gamopetalae
- Heteromerae
- Ebenales
- Styraceae

General characters

→ Trees, shrubs
→ Tomentose branches
→ Leaves - alternate, oblong, surface tomentose
→ Inflorescence - compound racemose
→ Calyx - 5 toothed, corolla 5 parted, gray; fused Stamens - 10, filaments connate at base into a short tube.
→ Fruit - dry capsule, sometimes winged, less often a fleshy drupe
→ One or 2 seeds

Genera included under Styraceae

- Alniphyllum
- Bruinsmia
- Changiostyrax
- Halesia
- Huodendron
- Melliodendron
- Parastyrax
- Pterostyrax
- Rehderodendron
- Sinojackia
- Styrax

Sub class: Gamopetalae

Series: Bicarpellatae

Order: gentianales

Families:

Oleaceae

Salvadoraceae

Apocyanaceae

Asclepiadaceae

Loganiaceae

Gentianaceae

Families of Dicotyledons

Oleaceae

Classification (Bentham and Hooker)

Phanerogams
Dicotyledons
Gamopetalae
Bicarpellatae
Gentianales
Oleaceae

General characters

→ Trees or shrubs, sometimes leanas, scandent or twining.

→ Leaves – opposite, simple or compound, exstipulate. Pellate hairs occur on the leaf epidermis. In *Jasminum*, the leaf axil subtends several buds arranged one above the other, these buds are accessory buds. In *Syringa*, the scale of the winter bud represents renderdeveloped leaf and is considered as leaf sheath phyllode.

→ Inflorescence – Terminal or Axillary cymes, Panicles or sometimes racemes. *Jasminum* – dichasium, *Syringa* – Panicle.

→ Flowers are bisexual, hypogynous, rarely unisexual; Actinomorphic, with small pedicel.

→ Calyx consists of 4 sepals, valvate, rarely absent; *Fraxinus* - sepals and petals both are absent.

→ Corolla consists of 4 or 5 petals, connate in Salver-shaped or other forms or rarely free, imbricate mostly but sometimes valvate, rarely absent;

→ Androecium consists of 2 stamens, sometimes 4, anther dithecous; distinct, often apiculate by extension of the connective, dehiscing longitudinally.

→ Gynoecium consists of 2 carpels, 1 pistil, Superior ovary, 2 locules, axile placentation; 2 ovules in each locule; Ovule anatropous, ascending or pendulous, Style 1 or absent, simple; Stigmas 1-2, bilobed.

→ Fruit— Berry or Drupe or loculicidal capsule or circumscissile capsule or Samara

→ Seeds - with or without endosperm; Endosperm when present is fleshy; embryo straight

Genera included under Oleaceae

- Abeliophyllum
- Chionanthus
- Comoranthus
- Fontanesia
- Forestiera
- Forsythia
- Fraxinus
- Haenianthus
- Hesperelaea
- Jasminum
- Ligustrum
- Linociera
- Menodora
- Myxopyrum
- Nestegis
- Noronhia
- Notelaea
- Nyctanthes
- Olea
- Osmanthus
- Phyllyrea
- Picconia
- Schrebera
- Syringa
- Tessarandra

Chapter-105
Salvadoraceae
Classification (Bentham and Hooker)
- Phanerogams
- Dicotyledons
- Gamopetalae
- Bicarpellatae
- Gentianales
- Salvadoraceae

General characters
→ Trees or shrubs; sometimes spinose with axillary spines; xerophytic

→ Leaves - opposite, somewhat fleshy or herbaceous or leathery; shortly petiolate; simple; lamina - entire; leaves stipulate or exstipulate; Margins entire.

→ Plants hermaphrodite or dioecious.

→ Inflorescence - Panicles or fascicles.

→ Flowers - regular, tetramerous; tetracyclic. Hypogynous disk present (intrastaminal) or absent

→ Perianth with distinct calyx and corolla.

→ Calyx consists of 2-5 sepals in 1 whorl, gamosepalous, blunt lobed; regular, imbricate or valvate

→ Corolla consists of 4-5 petals; 1 whorled; Poly or gamopetalous; corolla lobes markedly longer than the tube. Corolla imbricate or contorted, regular.

→ Androecium - 4-10 stamens; 1 or 2 whorled; Stamens isomerous with perianth; Anthers - dorsifixed dehiscing via longitudinal slits.

→ Gynoecium - bicarpellary, syncarpous, superior. Ovary 1 or 2 locular; Style-1, apical; Stigma - 1 or 2 entire or lobed; placentation basal (unilocular, or axile (bilocular). 1-2 ovules per locule.

→ Fruit - fleshy, indehiscent berry or drupe.
→ Seeds - non-endospermic.

Genera included under Salvadoraceae
→ Azima   → Dobera   → Salvadora

Apocynaceae

Classification (Bentham and Hooker)

- Phanerogams
- Dicotyledons
- Gamopetalae
- Bicarpellatae
- Gentianales
- Apocynaceae

General Characters

→ herbs, erect or twining shrubs or trees. Species of _Landolphia_ and _Clitandra_ are climbing shrubs.
→ Latex is present in most of the genera
→ Branched tap root
→ Stem - erect, branched, solid, glabrous rarely tuberlike and thick.
→ Leaves - simple, petiolate, usually opposite decussate. Rarely - alternate or whorled (_Nerium_). Usually exstipulate rarely stipulate
→ Inflorescence - usually cymose; rarely solitary (_vinca_), corymbose cymes (_Carissa_), terminal cymes (_Plumeria_), umbellate branched panicled cymes (_Alstonia_)
→ Flowers - pedicellate, bracteate, bracteolate, hermaphrodite, actinomorphic, regular, sometimes

slightly zygomorphic, complete, hypogynous and pentamerous. Rarely tetramerous with reduction to two in the pistil.

→ Calyx – 5 sepals, gamosepalous; calyx is generally divided almost to the base. Aestivation – quincuncial.

→ Usually corolla consists of 5 petals, gamopetalous; It is generally salver or funnel shaped. Corolla tube usually possesses hairy appendage or scales. Aestivation contorted.

→ Androecium consists of 5 stamens alternating with the petals. Stamens are situated on the tube or the throat of the corolla (epipetalous). Filaments – short, anthers introrse, polyandrous or connate and often adhere to the stigma. Antherlobes are sometimes empty at their base and prolonged into spines.

→ Gynoecium – bicarpellary, apocarpous or syncarpous, superior, sometimes partly inferior (Plumeria). Style simple, stigma thick often bilobed; usually a nectar secreting disc is situated beneath the gynoecium; Ovary superior to half inferior; Syncarpous – placentation parietal or axile; apocarpous – placentation marginal.

→ Fruit – follicles, berry, coconut like (Cerbera) – drupe, two valved capsule.

→ Dry fruits, seeds are generally winged or bear tuft of hairs at base; embryo straight; Pollination – insects.

Genera included under Apocynaceae

- Acokanthera
- Adenium
- Aganonerion
- Aganosma
- Alafia
- Allamanda
- Allomarkgrafia
- Allowoodsonia
- Alstonia
- Alyxia
- Amocalyx
- Ambelania
- Amsonia
- Ancylobotrys
- Anechites
- Angadenia
- Anodendron
- Apocynum
- Arduina
- Artia
- Asketanthera
- Aspidosperma
- Baissea
- Beaumontia
- Bousigonia
- Cabucala
- Callichilia
- Calocrater
- Cameraria
- Carissa
- Carpodinus
- Carruthersia
- Carvalhoa
- Cascabela
- Catharanthus
- Cerbera
- Cerberiopsis
- Chamaeclitandra
- Chilocarpus
- Chonemorpha
- Cleghornia
- Clitandra
- Condylocarpon
- Couma
- Craspidospermum
- Crioceras
- Cycladenia
- Cyclocotyla
- Cylindropsis
- Delphyodon
- Deweverella
- Dictyophleba
- Dipladenia
- Diplorhynchus
- Dyera
- Ecdysanthera
- Echites
- Elytropus
- Epigynum
- Eucorymbia
- Farquaharia
- Fernaldia
- Forsteronia
- Funtumia
- Galactophora
- Geissospermum
- Gonioma
- Grisseea
- Hancornia
- Haplophyton
- Himatanthus
- Holarrhena
- Hunteria

- → Hymenolophus
- → Ichnocarpus
- → Ixonema
- → Ixodonerium
- → Kamettia
- → Kibatalia
- → Kopsia
- → Lacmellea
- → Landolphia
- → Laubertia
- → Laxoplumeria
- → Lepinia
- → Lepiniopsis
- → Leuconotis
- → Lochnera
- → Lyonsia
- → Macoubea
- → Macropharynx
- → Macrosiphonia
- → Malouetia
- → Mandevilla
- → Mascarenhasia
- → Melodinus
- → Meschites
- → Mecrechtites
- → Microplumeria
- → Molongum
- → Mortonella
- → Motandra
- → Mucoa
- → Neobracea
- → Neocouma
- → Nerium
- → Nouettea
- → Ochrosia
- → Odontadenia
- → Oncinotis
- → Orthopichonia
- → Pachypodium
- → Pachouria
- → Papuechites
- → Parahancornia
- → Parameria
- → Pareepigynum
- → Parsonsia
- → Peltastes
- → Pentalinon
- → Petchia
- → Pleralima
- → Plectaneia
- → Pleiocarpa
- → Pleioceras
- → Plumeria
- → Pottsia
- → Prestonia
- → Pycnobotrya
- → Quiotania
- → Rauwolfia
- → Rhabdadenia
- → Rhazya
- → Rhigospira
- → Rhodocalyx
- → Rhyncodia
- → Saba
- → Salpinctes
- → Schizozygia
- → Secodatia
- → Sindechites
- → Skytanthus
- → Spirolobium
- → Spongiosperma
- → Stemmadenia
- → Stephanostegia
- → Stephanostemma
- → Stipecoma
- → Strempeliopsis
- → Strophanthus
- → Tabernaemontana
- → Tabernanthe
- → Temnadenia
- → Thenardia
- → Thevetia

- → Tintinnabularia
- → Trachelospermum
- → Urceola
- → Urnularia
- → Vahadenia
- → Vallariopsis
- → Vallaris
- → Vallesia
- → Vinca
- → Voacanga
- → Willughbeia
- → Woytkowskia
- → Wrightia
- → Xylinabaria
- → Xylinabariopsis.

Asclepiadaceae

Classification (Bentham and Hooker)

- Phanerogams
- Dicotyledons
- Gamopetalae
- Bicarpellatae
- Gentianales
- Asclepiadaceae

General Characters:

→ Mostly erect herbs or woody climbers but some are succulent. Most of the plants possess xerophytic characters.

→ Leaves - simple, sub sessile and exstipulate. Many cases - fleshy leaves covered with a coating of wax. (Calotropis); leaves are very much reduced and represented by spines and scales in *Stapelia*.

→ Inflorescence - cymose, racemose

→ Flowers - Pedicellate, bracteate, hermaphrodite, actinomorphic, rarely zygomorphic, complete. Pentamerous with 3 regularly alternating pentamerous whorls of calyx, corolla, androecium but the carpel number is reduced to 2 in gynoecium. Usually the flowers are small in size, rarely large sized flowers are present (Ceropegia)

→ Calyx consists of 5 sepals, which are either

Polysepalous or somewhat connate at the base with the odd petal posterior. Aestivation - quincuncial or rarely valvate

→ Corolla consists of 5 united petals, rotate but in Stephanotis the corolla tube is long, forming a salver shaped corolla. Ceropegia - corolla is pitcher-like in appearance. Aestivation - Contorted and rarely valvate. Very often these hairs or appendages are found inside or at the mouth of the corolla forming the corolline Corona.

→ Androecium consists of 5 stamens; anthers united laterally; giving rise to a five sided blunt cone, which is usually attached to the inside of the stigma head. With the result of this union of anthers and pistils, gynostegium is formed. Pollen grains of each half anther associated in tetrads are found in a sac like structure known as Pollinium. Anthers almost sessile. In the members of the family translator consists of 2 parts — (1) Corpusculum and (2) a pair of arms. The corpusculum is attached to the margin of stigma head between the anthers and the pair of arms is attached to the pollinia of the adjacent anther halves.

Certain appendages or horn like projections are given out from the backs of the anthers, are known as Cuculi, which secrete and store honey. Appendages form a corona known as staminal corona

→ Gynoecium consists of 2 carpels; ovaries remain free, but styles unite to form a common swollen stigma head. This stigma head may be flat or somewhat conical and sometimes even beaked. Receptive surfaces of the stigma are situated on the edge or on the underside of it. Ovary superior; Placentation – Marginal with many ovules

→ Fruit – Pair of follicles
→ Seeds are flat, ovate oblong and are crowned by a tuft of hairs. These hairs facilitate the dispersal of the seeds by wind. Embryo – large.

Genera included under Asclepiadaceae

- Absolmsia
- Adelostemma
- Aidomene
- Amblyopetalum
- Amblystigma
- Anatropanthus
- Anisopus
- Anisotoma
- Anomotassa
- Asaugia
- Asclepias
- Aspidoglossum
- Astephanus
- Barjonia
- Belostemma
- Bidaria
- Biondia
- Blepharodon
- Blyttia
- Brachystelma
- Calotropis
- Campestigma
- Caralluma
- Ceropegia
- Cibirhiza
- Cleonura
- Clemensiella
- Cononietta
- Cordylogyne
- Corollonema
- Cosmostigma
- Costantina
- Cyathostelma
- Cynanchum
- Dactylostelma
- Dalzielia
- Decabelone
- Decanema
- Decanemopsis
- Dicarpophora
- Diplolepis
- Diplostigma
- Dischidanthus
- Dischidia
- Ditassa
- Dittoceras
- Dolichopetalum
- Dolichostegia
- Dorystephania
- Dregea
- Drepanostemma
- Duvalia
- Duvaliandra
- Echidnopsis
- Edithcolea
- Emicocarpus
- Emplectanthus
- Eustegia
- Fanninia
- Fischeria
- Fockea
- Folotsia
- Frerea
- Funastrum
- Genianthus
- Glossonema
- Glossostelma
- Gomphocarpus
- Gongronema
- Goniaonthelma
- Goniostemma
- Gonolobus
- Graphistemma
- Gunnessia
- Gymnema
- Gymnemopsis
- Harmandiella

→ Hemipogon
→ Heterostemma
→ Heynella
→ Hickenia
→ Holostemma
→ Hoodia
→ Hoodeopsis
→ Hoya
→ Hoyella
→ Huernia
→ Huerniopsis
→ Hypolobus
→ Ischnostemma
→ Jacaima
→ Janakia
→ Jobinia
→ Kanahia
→ Karimbolea
→ Kerbera
→ Labidostelma
→ Lagoa
→ Lavrania
→ Leichardtia
→ Leptadenia
→ Lhotzkyella
→ Lugonia
→ Lygisma
→ Macroditassa
→ Macropetalum
→ Macroscepis
→ Mahafalia
→ Mahawoa
→ Manothrix
→ Margaretta
→ Marsdenia
→ Matelea
→ Melinia
→ Meresaldia
→ Merrillanthus
→ Metaplexis
→ Metastelma
→ Micholitzea
→ Microdactylon
→ Microloma
→ Microstelma
→ Miraglossum
→ Mitostigma
→ Morrenia
→ Nautonia
→ Nematostemma
→ Neoschumannia
→ Nephradenia
→ Notechidnopsis
→ Odontanthera
→ Odontostelma
→ Oncinema
→ Oncostemma
→ Ophionella
→ Orbea
→ Orbeanthus
→ Orbeopsis
→ Oreosparte
→ Orthanthera
→ Orthosia
→ Oxypetalum
→ Pachycarpus
→ Pachycymbium
→ Papuastelma
→ Parapodium
→ Pectinaria
→ Pentabothra
→ Pentacyphus
→ Pentarrhinum
→ Pentaschme
→ Pentastelma
→ Pentatropis
→ Peplonia
→ Pergularia
→ Periglossum
→ Petalostelma
→ Petopentia

- Pherotrichis
- Piaranthus
- Platykeleba
- Pleurostelma
- Podandra
- Podostelma
- Prosopostelma
- Pseudolithos
- Ptycanthera
- Pycnoneurum
- Pycnorhachis
- Quaqua
- Quisumbingia
- Raphistemma
- Rhyncharrhena
- Rhynchostigma
- Rhyssolobium
- Rhyssostelma
- Rhytidocaulon
- Riocreuxia
- Rojasia
- Sarcolobus
- Sarcostemma
- Schistogyne
- Schistonema
- Schizoglossum
- Schubertia
- Scyphostelma
- Secamone
- Secamonopsis
- Sehagiria
- Sesyrianthus
- Solenostemma
- Sphaerocodon
- Spirella
- Stapelia
- Stapelianthus
- Stapeliopsis
- Stathmostelma
- Steleostemma
- Stelmagonum
- Stelmatocodon
- Stenomeria
- Stenostelma
- Stigmatorhynchus
- Strobopetalum
- Stuckertia
- Swynnertonia
- Tassadia
- Tavaresia
- Telminostelma
- Telosma
- Tenaris
- Tetracustelma
- Tetraphysa
- Thozetia
- Toxocarpus
- Treutlera
- Trichocaulon
- Trichosacme
- Trichosandra
- Tridentea
- Tromotriche
- Tweedia
- Tylophora
- Tylophoropsis
- Vailia
- Vincetoxicopsis
- Vincetoxicum
- Voharanga
- Vohemaria
- White sloanea
- Widgrenia
- Woodia
- Xysmalobium

Chapter - 108
Loganiaceae

Classification (Bentham and Hooker)

- Phanerogams
- Dicotyledons
- Gamopetalae
- Bicarpellatae
- Gentianales
- Loganiaceae

General characters

→ Herbs or shrubs
→ Leaves – opposite, herbaceous or leathery or membranous; petiolate to sessile; commonly connate; simple; entire; linear or lanceolate or oblong or ovate; stipulate or exstipulate; stipules interpetiolar.
→ Plants hermaphrodite or dioecious
→ Female flowers with or without staminodes; In Male flowers – gynoecium pistillodial.
→ Perianth with distinct calyx and corolla.
→ Calyx usually 5, rarely 4; 1 whorled, gamosepalous, blunt lobed; regular, imbricate; Corolla 5 or 4; 1 whorled, gamopetalous (often internally hairy); imbricate; campanulate or

hypocrateriform or rotate ; regular
→ Androecium - 5 or 4 stamens, epipetalous, 1 whorled; stamens inserted midway down the corolla tube. Anthers - dorsifixed; dehiscing via longitudinal slits; introrse.
→ Gynoecium - bicarpellary, syncarpous, superior; ovary 2 locular; ovary sessile; styles-1, apical, stigma-1; lobed, clavate or capitate; placentation axile; ovules -1 to many per locule.
→ Fruit - non fleshy, dehiscent or schizocarp or capsule
→ Seeds - endospermic.

Genera included under Loganiaceae
→ Logania

Chapter - 109
Gentianaceae

Classification (Bentham and Hooker)

Phanerogams
Dicotyledons
Gamopetalae
Bicarpellatae
Gentianales
Gentianaceae

General characters

→ Herbs (often with dichotomous branching); autotrophic or parasite; Annual or biennial.
→ Leaves - opposite and decussate or rarely whorled or alternate, spiral; herbaceous or membranous; petiolate to sessile; simple; epulvinate; lamina - entire, margins - entire; exstipulate
→ Plants hermaphrodite; unisexual flowers are present in some cases.
→ Inflorescence - cymes; solitary, axillary flowers
→ Flowers - bracteate or ebracteate; bracteolate or ebracteolate; irregular; 4-12 merous; tetracyclic. Some genera show the floral receptacle developing a gynophore.

- Perianth with distinct calyx and corolla;
- Calyx consists of 4-12 sepals, 1 whorled, polysepalous but usually gamosepalous, entire, lobulate or blunt-lobed; regular; imbricate. Epicalyx present or absent.
- Corolla consists of 4-12 petals in 1 whorl, gamopetalous usually contorted or imbricate; campanulate or funnel shaped or cyathiform; regular; often showy.
- Androecium 4-12 stamens; free of one another, rarely coherent; 1 whorled; Anthers - dorsifixed, versatile or rarely basifixed, non versatile; introrse; dehisce via longitudinal slits or apical pores.
- Gynoecium - bicarpellary, syncarpous, superior. Ovary usually unilocular rarely bilocular; Style-1 apical; stigma 1 or 2; Placentation usually parietal, rarely free central; 1 - Many ovules per cavity.
- Fruit - non-fleshy dehiscent capsule (or) rarely fleshy, non dehiscent, berry.
- Seeds - copiously endospermic.

Genera Included under Gentianaceae

- Bartonia
- Belmontia
- Bisgoeppertia
- Blackstonia
- Canscora
- Celiantha
- Centaurium
- Chironia
- Chorisepalum
- Cicendia
- Comastoma
- Congolanthus
- Cotylanthera
- Coutoubea
- Cracosna
- Crawfurdia
- Curtia
- Deianira
- Djaloniella
- Enicostema
- Eustoma
- Exaculum
- Exacum
- Wurdackanthus
- Faroa
- Frasera
- Gentiana
- Gentianella
- Gentianopsis
- Gentianothamnus
- Halenia
- Hockinia
- Hoppea
- Irlbachia
- Ixanthus
- Jaeschkea
- Karina
- Latouchea
- Lehmanniella
- Lisianthius
- Lomatogoniopsis
- Lomatogonium
- Macrocarpaea
- Megacodon
- Microrphium
- Monodiella
- Neblinantha
- Neurotheca
- Zonanthus
- Obolaria
- Oreonesion
- Ornichia
- Orphium
- Parajaeschkea
- Prepusa
- Pterygocalyx
- Pycnosphaera
- Rogersonanthus
- Sabatia
- Schinzeella
- Schultesia
- Sebaea
- Senaea
- Sipapoantha
- Swertia
- Symbolanthus
- Symphyllophyt
- Tachia
- Tachiadenus
- Tapeinostemon
- Tetrapollinia
- Trepterospermum
- Urogentias
- Veratrilla
- Voyria
- Voyriella
- Zygostigma

Sub class: Gamopetalae

Series: Bicarpellatae

Order: polemoniales

Families:

Polemoniaceae

Hydrophyllaceae

Boragineae

Convolvulaceae

Solanaceae

Chapter - 110
Polemoniaceae

Classification (Bentham and Hooker)

Phanerogams
Dicotyledons
Gamopetalae
Bicarpellatae
Polemoniales
Polemoniaceae

General characters

→ Herbs, shrubs or trees; Annual to perennial.
→ Leaves - alternate or opposite or whorled; when alternate, spiral; Petiolate to sessile; Simple or compound; entire, serrate or dentate margin; exstipulate.
→ Plants hermaphrodite.
→ Flowers - solitary; inflorescence - cymes or heads or corymbs.
→ Flowers - bracteolate or ebracteolate; regular; somewhat irregular (slightly zygomorphic); pentamerous; tetracyclic; Hypogynous disk present; Intrastaminal.
→ Perianth with distinct calyx and corolla
→ Calyx consists of 5 sepals, 1 whorled; gamosepalous,

five blunt-lobed; bilabiate to regular; imbricate or valvate
- → Corolla consists of 5 petals, 1 whorled, gamopetalous usually contorted, campanulate or funnel shaped, bilabiate to regular.
- → Androecium - 5 stamens, epipetalous; 1 whorled; anthers - dehiscing via longitudinal slits.
- → Gynoecium 2-5 carpelled; syncarpous; superior. Ovary 2-5 locular; ovary sessile; style -1, filiform. Stigma 2-5 lobed. Placentation axile; ovules 1-many per locule; anatropous.
- → Fruit - non-fleshy, dehiscent or indehiscent, capsule (loculicidal)
- → Seeds - endospermic.

Genera included under Polemoniaceae
- Acanthogilia
- Aliciella
- Allophyllum
- Bonplandia
- Cantua
- Collomia
- Eriastrum
- Gilia
- Gymnosteris
- Huthia
- Ipomopsis
- Langloisia
- Leptodactylon
- Linanthus
- Loeselia
- Loeseliastrum
- Navarretia
- Phlox
- Polemonium

Chapter-III
Hydrophyliaceae

Classification (Bentham and Hooker)

- Phanerogams
- Dicotyledons
- Gamopetalae
- Bicarpellatae
- Polemoniales
- Hydrophyliaceae

General characters

→ Annual or perennial herbs or shrubs
→ Prostrate or erect stem
→ Taproot
→ Flowers - bisexual, normally radial, with 5 petals and 5 stamens
→ Stamens attached to the base of the petals
→ Ovary superior; 2 rarely 4 carpels, syncarpous; forming a single chamber.
→ Fruit - capsule

Genera included under Hydrophyliaceae

→ Codon
→ Draperia
→ Ellisia
→ Emmeranthe
→ Eriodictyon
→ Eucrypta

- → Hesperochiron
- → Hydrophyllum
- → Lemmonia
- → Miltitzia
- → Nama
- → Nemophila
- → Phacelia
- → Pholistoma
- → Romanzoffia
- → Tricardia
- → Turricula
- → Wigandia

Chapter 112: Boraginaceae

Classification (Bentham and Hooker)

    Phanerogams
    Dicotyledons
    Gamopetalae
    Bicarpellatae
    Polemoniales
      Boraginaceae

General characters:

→ Annual or perennial herbs, shrubs or rarely trees. Main features of the family is the presence of thick walled stiff hairs in the herbaceous species, usually scabrous or hispid hairy and sometimes glabrous herbs.

→ Leaves — alternate, lowermost leaves sometimes opposite, simple, usually entire, exstipulate. In some species (*cynoglossum*), heterophylly is marked where the radical leaves differ from the cauline leaves; the cystoliths present in the leaves.

→ Inflorescence — scorpioid cyme or helicoid cyme that uncoil as the flowers open. Sometimes loosely cymose or with solitary flowers.

→ Flowers - bisexual, actinomorphic, rarely zygomorphic (Lycopsis), hypogynous

→ Calyx consists of 5 sepals, free or basally connate, imbricate or rarely valvate, sometimes irregular

→ Corolla consists of 5 petals, united, imbricate or contorted in bud, rotate, salverform, funnelform or campanulate;

→ Androecium consists of 5 stamens, epipetalous, equal or rarely unequal, dorsally appendaged (Borago); alternate with corolla lobes; anthers dithecous (2 celled), dehiscing longitudinally, basifixed or basally dorsifixed, introrse.

→ Gynoecium - bicarpellary, syncarpous, superior ovary. Placentation axile, rarely central or parietal; style-1, gynobasic or terminal, simple; stigma - simple, capitate or bilobed; ovules - anatropous or amphitropous; many or 2 in each locule

→ Fruit - loculicidal capsule or fruit of 4 nutlets or a 1-4 seeded nut or drupe

→ Seeds - minute, usually without endosperm or the endosperm is scanty and fleshy (Heliotropium)

Genera included under Boraginaceae

- Actinocarya
- Adelocaryum
- Afrotysonia
- Alkanna
- Amblynotus
- Amphibologyne
- Amsinckia
- Anchusa
- Ancistrocarya
- Anoplocaryum
- Antiotrema
- Antiphytum
- Arnebia
- Asperugo
- Auxemma
- Borago
- Bothriospermum
- Brachybotrys
- Brunnera
- Buglossoides
- Caccinia
- Caramona
- Cerinthe
- Chionocharis
- Choriantha
- Craniospermum
- Cryptantha
- Cynoglossopsis
- Cynoglossum
- Cynoglottis
- Cystostemon
- Dasynotus
- Decalepidanthus
- Echiochilon
- Echiostachys
- Echium
- Elizaldia
- Embadium
- Eritrichium
- Euploca
- Gastrocotyle
- Gyrocaryum
- Hackelia
- Halacsya
- Heliocarya
- Heliotropium
- Heterocaryum
- Huynhia
- Ivanjohnstonia
- Ixorhea
- Lacaitaea
- Lappula
- Lasiarrhenum
- Lasiocaryum
- Lepechiniella
- Lepidocordia
- Lindelophia
- Lithodora
- Lithospermum
- Lobostemon
- Macromeria
- Macrotomia
- Maharanga
- Mairetis
- Mattiastrum
- Mertensia
- Metaeritrichium
- Microcaryum
- Microula
- Mimophytum
- Moltkia

- → Moltkeopsis
- → Moritzia
- → Myosotideum
- → Myosotis
- → Neatostema
- → Nesocaryum
- → Nogalia
- → Nomosa
- → Nonea
- → Ogastemma
- → Omphalodes
- → Omphalolappula
- → Omphalotrigonotis
- → Onosma
- → Onosmodium
- → Oxyosmyles
- → Paracaryum
- → Pardoglossum
- → Patagonula
- → Pectocarya
- → Pentaglottis
- → Perittostema
- → Plagiobothrys
- → Pseudomertensia
- → Psilolaemus
- → Pteleocarpa
- → Pulmonaria
- → Rindera
- → Rochefortia
- → Rochelia
- → Rotula
- → Saccellium
- → Scapicephalus
- → Selkirkia
- → Sericostoma
- → Sinojohnstonia
- → Solenanthus
- → Stenosolenium
- → Stephanocaryum
- → Suchtelenia
- → Symphytum
- → Thaumatocaryum
- → Thyrocarpus
- → Tianschaniella
- → Tiquilia
- → Tournefortia
- → Trachelanthus
- → Trachystemon
- → Trichodesma
- → Trigonocaryum
- → Trigonotis
- → Ulugbekia
- → Valentiniella

Chapter - 113
Convolvulaceae

Classification (Bentham and Hooker)

    Phanerogams
    Dicotyledons
    Gamopetalae
    Bicarpellatae
    Polemoniales
    Convolvulaceae

General characters:

→ Usually twining herbs or shrubs, sometimes erect, rarely trees. Plants may be xerophytes, hydrophytes and sometimes parasites. When the plants are shrubby they produce thorns.

→ Roots - variable; long and slender where they serve the purpose of vegetative propagation; thick, fleshy, tuberous (*Ipomoea batatas*) where they store plenty of food. *Cuscuta reflexa*, root is represented by small haustoria meant for absorbing food from the host.

→ Inflorescence - dichasial cyme or solitary axillary. Sometimes - racemose or panicle.

— Leaves are simple, exstipulate, usually petiolate and alternate. *Cuscuta*, leaves are reduced to scales or altogether absent.

→ Flowers – bracteate, bracteolate, usually pedicellate bisexual rarely unisexual by abortion, actinomorphic usually showy, hypogynous, usually pentamerous (except gynoecium). Bracts arranged in pairs and form an involucre. Sometimes flowers are cleistogamous (*Dichondra*)

→ Calyx consists of 5 sepals, polysepalous but sometimes united at the base. It is persistent and imbricate.

→ Corolla is 5 lobed or entire, usually campanulate, funnel shaped or salver shaped. Corolla is appendaged in *cuscuta*. Aestivation – valvate rarely imbricate.

→ Androecium consists of 5 stamens, free, epipetalous at the base of corolla tube, alternate with petals; Anthers – dithecous, dorsifixed, dehisce by longitudinal slits, introrse, usually cup-shaped or ring like intrastaminal disc present

→ Gynoecium – bicarpellary, syncarpous, 1 pistil, ovary superior, Placentation – axile, bilocular with 1 to 2 ovules in each locule. Ovules anatropous; Style – single, simple and filiform. Stigma – terminal and capitate, sometimes 2 distinct stigma

→ Fruit – loculicidal capsule

→ Seed – endospermic, smooth or hairy with curved embryo. Pollination – entomophilous.

Genera included under convolvulaceae

- Aniseia
- Argyreia
- Astripomoea
- Blinkworthia
- Bonamia
- Breweria
- Calycobolus
- Calystegia
- Cardiochlamys
- Cladistigma
- Convolvulus
- Cordisepalum
- Cressa
- Decalobanthus
- Dichondra
- Dicranostyles
- Dinetus
- Dipteropeltis
- Erycibe
- Evolvulus
- Falckia
- Hewittia
- Hildebrandtia
- Hyalocystis
- Ipomoea
- Iseia
- Itzaea
- Jacquemontia
- Lepistemon
- Lepistemonopsis
- Lysiostyles
- Maripa
- Merremia
- Metaporana
- Nephrophyllum
- Neuropeltis
- Neuropeltopsis
- Odonellia
- Operculina
- Paralepistemon
- Pentacrostigma
- Pharbitis
- Polymeria
- Porana
- Poranopsis
- Rapona
- Rivea
- Sabaudiella
- Seddera
- Poranopsis
- Rapona
- Rivea
- Sabaudiella
- Seddera
- Stictocardia
- Stylisma
- Tetralocularia
- Tridynamia
- Turbina
- Wilsonia
- Xenostegia

Chapter – 114
Solanaceae

Classification (Bentham and Hooker)

Phanerogams
Dicotyledons
Gamopetalae
Bicarpellatae
Polemoniales
Solanaceae

General Characters

→ Either erect or climbing herbs, shrubs or small trees.
→ Taproot or adventitious root
→ Leaves- alternate or becoming opposite at or near the inflorescence, rarely whorled, exstipulate, entire, simple, rarely dissected or pinnate
→ Inflorescence – cymose. usually a typical axillary cyme or combination of cymes, rarely helicoid.
→ Flowers- usually pedicellate, hermaphrodite, regular (actinomorphic), slightly zygomorphic, hypogynous complete, bracts and bracteoles absent.
→ Calyx consists of 5 sepals, gamosepalous (limb usually 5-lobed or toothed), persistent and often much enlarged in fruit.

→ Corolla consists of 5 petals, gamopetalous, funnel shaped, rotate to tubular, corolla of various shapes but rarely 2 lipped, usually plicate or convolute (rarely valvate)

→ Androecium consists of 5 stamens, rarely 4 or 2. When the stamens are less than 5 the lost stamen is usually represented by staminoder. Stamens epipetalous and alternate, petals usually not equal in size. Anthers are dithecous, introrse dehisce by longitudinal slits or by apical pores, connective sometimes enlarged, usually hypogynous disc present which is quite apparent.

→ Gynoecium - bicarpellary, syncarpous, ovary superior bilocular, rarely tetralocular due to false septum, rarely 3-5 celled ovary, Axile placentation, ovules usually many on prominently peltate placentas, anatropous or somewhat amphitropous. Carpels placed obliquely in the flowers. Single style, stigma bilobed and capitate

→ Fruit - berry or septicidal capsule

→ Seed - Albuminous, smooth or pitted, curved embryo

Genera included under Solanaceae

- Acnistus
- Anisodus
- Anthocercis
- Anthotroche
- Archiphysalis
- Althenaea
- Atropa
- Atropanthe
- Benthamiella
- Bouchetia
- Brachistus
- Browallia
- Brugmansia
- Brunfelsia
- Calibrachoa
- Capsicum
- Cestrum
- Chamaesaracha
- Combera
- Crenidium
- Cuatresia
- Cyphanthera
- Cyphomandra
- Datura
- Deprea
- Discopodium
- Duboisia
- Dunalia
- Dyssochroma
- Ectozoma
- Exodeconus
- Faliana
- Grabowskia
- Grammosolen
- Hawkesiophyton
- Heteranthia
- Hunzikeria
- Hyoscyamus
- Iochroma
- Jaborosa
- Jaltomata
- Juanulloa
- Latua
- Leptoglossis
- Leucophysalis
- Lycianthes
- Lycium
- Lycopersicon
- Mandragora
- Margaranthus
- Markea
- Melananthus
- Mellissia
- Merinthopodium
- Metternichia
- Nectouxia
- Nicandra
- Nicotiana
- Nierembergia
- Nothocestrum
- Oryctes
- Pantacantha
- Parabouchetia
- Pauia
- Petunia
- Phrodus
- Physalis
- Physochlaina
- Plowmania
- Protoschwenckia
- Przewalskia
- Scopolia

- → Sessea
- → Sesseopsis
- → Solandra
- → Solanum
- → Streptosolen
- → Symonanthus
- → Trianaea
- → Triguera
- → Tubocapsicum
- → Vassobia
- → Vestia
- → Withania
- → Witheringia

Sub class: Gamopetalae

Series: Bicarpellatae

Order: personales

Families:

Scrophularineae

Orobanchaceae

Lentibularieae

Columelliaceae

Gesneraceae

Bignoniaceae

Pedalineae

Acanthaceae

Chapter -15
Scrophulariaceae

Classification (Bentham and Hooker)

- Phanerogams
- Dicotyledons
- Gamopetalae
- Bicarpellatae
- Personales
- Scrophulariaceae

General characters

→ Shrubs, very rarely small trees; some are lianas, parasites, saprophytes, climbers
→ Branched Tap root system
→ Stem - erect, prostrate, creeping or ascending; herbaceous or woody; solid, branched
→ Leaves - alternate, opposite or whorled; simple, exstipulate, entire or pinnately lobed. In the aquatic genera (Ambulia) possess heterophylly. In such cases, the aerial leaves are entire and submerged and much dissected.
→ Inflorescence - Solitary (Linaria) or arranged in terminal spikes or racemes. Bracts and bracteoles are generally present. Bracts are brightly coloured in some cases. (Castilleja)

- → Flowers bisexual, zygomorphic, sometimes actinomorphic, hypogynous
- → Calyx consists of 5 sepals but in Euphrasia & Veronica, the posterior sepal is suppressed. <u>Calceolaria</u> 2 anterior sepals are completely united
- → Corolla consists of 5 lobes, gamopetalous, zygomorphic Imbricate aestivation.
- → Androecium consists of 4 stamens, didynamous and epipetalous. Anthers bicelled, appendages or hairy structures may be found, introrse, dehisce by longitudinal slit.
- → Gynoecium consists of 2 carpels, syncarpous, suited on nectar secreting disc. Ovary superior, bilocular, rarely unilocular, carpels median, and not obliquely placed. Style simple and stigma bilobed
- → Fruit - capsule rarely berry
- → Seed - small and numerous. Endospermic with straight or curved embryo.
- → Pollination entomophilous.

Genera included under Scrophulariaceae

- Acanthorrhinum
- Achetaria
- Adenosma
- Agathelpis
- Albraunia
- Alectra
- Allocalyx
- Alonsoa
- Analophyllon
- Amphianthus
- Amphidanthus
- Anarrhinum
- Anastrabe
- Angelonia
- Antherothamnus
- Anticharis
- Antirrhinum
- Aptosimum
- Aragoa
- Artanema
- Asarina
- Aureolaria
- Bacopa
- Bampsia
- Basistemon
- Baumia
- Benjaminia
- Besseya
- Bowkeria
- Brachystigma
- Brandisia
- Brookea
- Bryodes
- Bungea
- Buttonia
- Bythophyton
- Calceolaria
- Camptoloma
- Campylanthus
- Capraria
- Celsia
- Centranthera
- Centrantheropsis
- Chaenorhinum
- Charadrophila
- Cheilophyllum
- Chelone
- Chenopodeopsis
- Chionohebe
- Chionophila
- Clevelandia
- Cochlidiosperma
- Collinsia
- Colpias
- Conobea
- Craterostigma
- Crepidorhopalon
- Cromidon
- Cycniopsis
- Cycnium
- Cymbalaria
- Cyrtandromoea
- Dasutonia
- Deinostema
- Dermatobotrys
- Detzneria
- Diascia
- Diclis
- Digitalis
- Dintera

- → Diplacus
- → Dischisma
- → Dizygostemon
- → Dodartia
- → Dopatrium
- → Elacholoma
- → Encopella
- → Epixiphium
- → Eremogeton
- → Erinus
- → Escobedia
- → Esterhazya
- → Faxonanthus
- → Fonkia
- → Freylinia
- → Galvezia
- → Gambelia
- → Geochorda
- → Gerardiina
- → Ghikaea
- → Glekia
- → Globulariopsis
- → Glossostigma
- → Glumicalyx
- → Gosela
- → Graderia
- → Gratiola
- → Halleria
- → Harveya
- → Hebe
- → Hebenstretia
- → Hedbergia
- → Hemianthus
- → Hemiarrhena
- → Hemichaena
- → Hemimeris
- → Hemiphragma
- → Hiernia
- → Holmgrenanthe
- → Holzneria
- → Howelliella
- → Hydranthelium
- → Hydrotriche
- → Hygea
- → Hyobanche
- → Idefonsia
- → Isoplexis
- → Ixianthus
- → Jamesbrittenia
- → Jerdonia
- → Jovellana
- → Kashmiria
- → Keckiella
- → Kichxia
- → Lafuentea
- → Lagotis
- → Lamourouxia
- → Lancea
- → Legazpia
- → Leptorhabdos
- → Leucocarpus
- → Leucophyllum
- → Leucosalpa
- → Leucospora
- → Limnophila
- → Limosella
- → Linaria
- → Lindenbergia
- → Lindernia
- → Lophospermum
- → Lyperia
- → Mabrya
- → Macranthera
- → Maeviella
- → Magdalenaea
- → Manulea

- → Manuleopsis
- → Maurandella
- → Maurandya
- → Mazus
- → Mecardonia
- → Melanospermum
- → Melasma
- → Melosperma
- → Micranthemum
- → Micrargeria
- → Micrargeriella
- → Microcarpaea
- → Microdon
- → Mimetanthe
- → Mimulicalyx
- → Mimulus
- → Misopates
- → Mohavea
- → Monochasma
- → Monopera
- → Monttea
- → Moscheovia
- → Nemation
- → Nathaliella
- → Nemesia
- → Neogaerrrhinum
- → Neopicrorhiza
- → Nothochelone
- → Nothochilus
- → Nuttallanthus
- → Odecardes
- → Oftia
- → Omphalotrix
- → Ophiocephalus
- → Oreosolen
- → Otacanthus
- → Ourisia
- → Paederota
- → Paederotella
- → Parahebe
- → Parastriga
- → Paulownia
- → Peliostomum
- → Pennellianthus
- → Penstemon
- → Peplidium
- → Phygelius
- → Phyllopodium
- → Physocalyx
- → Picria
- → Picrorhiza
- → Preenanthus
- → Polycarena
- → Poroditia
- → Psammetes
- → Pseudobartsia
- → Pseudolysimachion
- → Pseudomelasma
- → Pseudorantium
- → Pseudosopubia
- → Pseudostriga
- → Pterygiella
- → Rehmannia
- → Rhamphicarpa
- → Raphispermum
- → Rhodochiton
- → Rhynchocorys
- → Russelia
- → Sacrocarpus
- → Schistophragma
- → Acherosepala
- → Schizotorenia
- → Schlegia
- → Schwalbea

- → Schweinfurthia
- → Scolophyllum
- → Scoparia
- → Scrofella
- → Scrophularia
- → Selago
- → Seymeria
- → Seymeriopsis
- → Shiuyinghua
- → Sibthorpia
- → Silviella
- → Siphonostegia
- → Sopubia
- → Spirostegia
- → Stemodia
- → Stemodiopsis
- → Stregina
- → Strobilopsis
- → Sutera
- → Synthyris
- → Teedia
- → Tetranema
- → Tetraselago
- → Tetraspidium
- → Tetraulacium
- → Thunbergianus
- → Tonella
- → Torenia
- → Tozzia
- → Triaenophora
- → Tricena
- → Trungboa
- → Tuerckheimocharis
- → Uroskinnera
- → Vellosiella
- → Verbascum
- → Veronica
- → Veronicastrum
- → Walafrida
- → Wightia
- → Wulfenia
- → Wulfeniopsis
- → Xizangia
- → Xylocalyx
- → Zaluzianska
- →:

Chapter-116
Orobanchaceae

Classification (Bentham and Hooker)

    Phanerogams
    Dicotyledons
    Gamopetalae
      Bicarpellatae
      Personales
      Orobanchaceae

General characters:

→ Achlorophyllous herbs
→ Leaves much reduced (to scales) or absent. Plants rather succulent; Parasitic; Annual, biennial or perennial; commonly rhizomatous or tuberous.
→ Leaves – small, alternate, spiral, Membranous; sessile, non-sheathing; simple; lamina – entire; lanceolate or oblong or ovate; exstipulate;
→ Plants hermaphrodite; Pollination entomophilous.
→ Inflorescence – Racemes or spikes.
→ Flowers – bracteate, very irregular; tetracyclic. Hypogynous disk present (fleshy).
→ Perianth with distinct calyx and corolla.
→ Calyx 4-5 sepals or rarely 2; 1 whorled; gamosepalous; blunt-lobed, or toothed; valvate

→ Corolla 5 petals, 1 whorled; gamopetalous; imbricate; persistent.

→ Androecium – 4 or 5 stamens, epipetalous; 1 whorled; Anthers – introse; dehisce via longitudinal slits.

→ Gynoecium – bi or tri carpellary; syncarpous; superior. Styles – 1, apical; stigma – 2–4, lobed; Placentation parietal; ovules 12 – many per cavity; anatropous.

→ Fruit – non-fleshy, dehiscent, capsule (loculicidal or valvular)

→ Seeds – endospermic.

Genera included under Orobanchaceae

- Aeginetia
- Boschniakia
- Buchnera
- Chritisonia
- Cistanche
- Conopholis
- Epifagus
- Gleadovia
- Kopsiopsis
- Lathraea
- Mannafettaea
- Necranthus
- Orobanche
- Phacellanthus
- Phelypaea
- Platypholis
- Radamaea
- Xylanche

Chapter-118
Columelliaceae

Classification (Bentham and Hooker)

Phanerogams
Dicotyledons
Gamopetalae
Bicarpellatae
Personales
Columelliaceae

General Characters

→ Bitter trees or shrubs
→ Leaves - persistent, opposite, simple, entire, exstipulate; Margin - entire or dentate
→ Plants hermaphrodite
→ Flowers solitary or in inflorescence - cymes.
→ Flowers - bracteolate, irregular (zygomorphic), tetracyclic; free hypanthium present
→ Perianth with distinct calyx and corolla
→ Calyx consists of 4-8 sepals in 1 whorl, gamo or polysepalous; slightly imbricate or valvate.
→ Corolla consists of 4-8 petals, 1 whorled; gamopetalous. Corolla lobes markedly longer

than tube; imbricate; sub-irregular, yellow.
→ Androecium - 2 stamens, 1 whorled, oppositi sepalous; Anthers connivent with a broad connective; dehiscing via longitudinal slits.
→ Gynoecium - bicarpellary, syncarpous, inferior. Ovary 1 locular. Styles - 1, short, thick and apical; Stigma - 2 or 4 lobed. Placentation Parietal. Ovules Many per cavity; anatropous.
→ Fruit - non-fleshy, dehiscent, capsule (septicidal or valvular)
→ Seeds - endospermic.

Genera included under columelliaceae
→ columellia

Chapter-119
Gesneriaceae

Classification (Bentham and Hooker)

Phanerogams
Dicotyledons
Gamopetalae
Bicarpellatae
Personales
Gesneriaceae

General characters

→ Bilaterally symmetrical, bisexual flowers;
→ 2 lipped corolla of five fused petals; five-lobed calyx (sepals); 2 or 4 rarely 5 anthers that are lightly joined together or in pairs
→ Ovary superior or partly inferior
→ Placentation Parietal
→ Ovary 1 loculed
→ Seeds - numerous, small
→ Herbaceous or slightly woody plants

Genera included under Gesneriaceae

→ Achimenes
→ Aeschynanthus
→ Agalmyla
→ Alloplectus
→ Alsobia
→ Amalophyllon
→ Ancylostemon
→ Anna
→ Anodiscus

- → Asteranthera
- → Besleria
- → Boea
- → Briggsia
- → Bucinellina
- → Capanea
- → Chirita
- → Chiritopsis
- → Chrysothemis
- → Codonanthe
- → Codonoboea
- → Columnea
- → Conandron
- → Corallodiscus
- → Corytoplectus
- → Crantzia
- → Cyrtandra
- → Depanthus
- → Deinostigma
- → Diastema
- → Didissandra
- → Didymocarpus
- → Drymonia
- → Episcia
- → Eucodonia
- → Fieldia
- → Gasteranthus
- → Gesneria
- → Glossoloma
- → Gloxinella
- → Gloxinia
- → Gloxiniopsis
- → Haberlea
- → Hemiboea
- → Henckelia
- → Isometrum
- → Jancaea
- → Jerdonia
- → Koellikeria
- → Kohleria
- → Lenbrassia
- → Loxocarpus
- → Loxostigma
- → Lysionotus
- → Mandirola
- → Microchirita
- → Mitraria
- → Monophyllaea
- → Monopyle
- → Nepeanthus
- → Nautilocalyx
- → Negria
- → Nematanthus
- → Neomortonia
- → Niphaea
- → Nomopyle
- → Opithandra
- → Oreocharis
- → Paliavana
- → Paraboea
- → Paradrymonia
- → Pearcea
- → Petrocodon
- → Petrocosmea
- → Phinaea
- → Primulina
- → Ramonda
- → Raphiocarpus
- → Reldia
- → Rhabdothamnus
- → Rhynchoglossum
- → Rhytidophyllum
- → Sanango
- → Sarmienta
- → Saintpaulia
- → Seemannia
- → Sinningia

- → Smithiantha
- → Solenophora
- → Sphaerorrhiza
- → Stauranthera
- → Streptocarpus
- → Titanotrichum
- → Tremacron

Chapter 117: Lentibulariaceae

Classification (Bentham and Hooker)

    Phanerogams
        Dicotyledons
            Gamopetalae
                Bicarpellatae
                    Personales
                        Lentibulariaceae

General Characters:

→ Annual and Perennial herbs; aquatic often rootless; or plants found in wet or marshy places, predominantly insectivorous.

→ Leaves — alternate or in basal rosettes; often dimorphic in aquatic members with finely divided submerged leaves and bearing insectivorous bladders.

→ Various kinds of glands are found on the surface of leaf. In _Utricularia_, glands are small capitate gland; in _Pinguicula_ the glands are stalked and each gland has a basal cell, stalk cell and disc-shaped multicellular secretory cap; some simple glandular hairs are also present.

- Inflorescence – Racemose, racemes of spikes, scapes or solitary on scape
- Flowers – bracteate, pedicels each with a pair of bracteoles, bisexual, zygomorphic and hypogynous.
- Calyx consists of 5 sepals, gamosepalous, united in open bilabiate or 5 lobed, usually persistent, valvate or sometimes imbricate
- Corolla consists of 5 petals, gamopetalous, united in bilabiate or personate corolla, lower lip spurred or saccate at base, with a palate very variable in form near the throat.
- Androecium consists of 2 stamens, arising from extreme base of corolla tube, anthers monothecous, anther may be sometimes medianly constricted, dehisce longitudinally
- Gynoecium – bicarpellary, syncarpous (united in 1 cell), superior ovary, style very short or stigma sessile, bilobed. Ovules many, anatropous, rarely only 2; Placentation - free central; ovules often sunken in placental tissue
- Fruit – Many seeded capsule; seeds – small, exendospermic; Pollination – entomophilous

Genera included under Lentibulariaceae
- Genlisea
- Pinguicula
- Polypompholyx
- Utricularia

Chapter-120
Bignoniaceae

Classification (Bentham and Hooker)

- Phanerogams
- Dicotyledons
- Gamopetalae
- Bicarpellatae
- Personales
- Bignoniaceae

General Characters

→ Mostly trees and shrubs; Shrubs are climbers by tendrils or twiners, rarely root climbers.

→ Anamolous secondary thickening in the stems is of universal occurrence. Here the cambium form xylem and phloem in very much varying quantities that is in some segments much greater amount of xylem is produced than phloem and in others more phloem than xylem. As a result a characteristic structure with ridged and furrowed xylem cylinder is formed.

→ Leaves — opposite, compound pinnate and exstipulate. Leaves 2-3 times pinnate

→ Tendrils — are the metamorphosed terminal leaflets of a compound pinnate leaf, hence the

tendrils are typical leaf tendrils.
- → Inflorescence — racemose or cymose inflorescence, either dichasium or helicoid cyme or panicle of dichasia or clustered.
- → Flowers — showy, zygomorphic, hermaphrodite, hypogynous.
- → Calyx consists of 5 sepals, united in 5 lobed or toothed, sometimes bilabiate irregular calyx, lobes imbricate rarely valvate. Spathaceous calyx is present in <u>Dolichandrone</u>.
- → Corolla consists of 5 petals, united in funnel-shaped or bell shaped (campanulate) or sometimes bilabiate corolla; Aestivation — descending imbricate.
- → Androecium — stamens 4, didynamous, posterior 5th stamen often reduced to staminode or absent, inserted on the basal part of corolla tube, anthers dithecous, anther lobes are placed little above the other; dehisce longitudinally.
- → Gynoecium — bicarpellary, syncarpous, carpels united in bilocular superior ovary rarely unilocular. Ovules many, anatropous, micropyle directed downward. Placentation axile rarely parietal. Style simple, filiform, stigma bilobed. Fleshy annular hypogynous disc is present around the ovary.
- → Pollination by small birds.
- → Fruit — capsule or berry like. Seeds — exendospermous, often winged, rarely comose, not winged.

Genera included under Bignoniaceae

- Adenocalymna
- Amphilophium
- Amphitecna
- Anemopaegma
- Argylia
- Astianthus
- Astianthus
- Barnettia
- Bignonia
- Callichlamys
- Campsidium
- Campsis
- Catalpa
- Catophractes
- Ceratophytum
- Chilopsis
- Clytostoma
- Colea
- Crescentia
- Cuspidaria
- Cybistax
- Delostoma
- Deplanchea
- Digomphia
- Dinklageodoxa
- Disticella
- Distictis
- Dolichandra
- Dolichandrone
- Eccremocarpus
- Ekmanianthe
- Fernandoa
- Fridericia
- Gardnerodoxa
- Glaziova
- Godmania
- Haplolobium
- Haplophragma
- Heterophragma
- Hieris
- Incarvillea
- Jacaranda
- Kigelia
- Lamiodendron
- Leucocalantha
- Lundia
- Macfadyena
- Macranthisiphon
- Manaosella
- Mansoa
- Markhamia
- Martinella
- Melloa
- Memora
- Millingtonia
- Mussatia
- Neojobertia
- Neosepicaea
- Newbouldia
- Nyctcalos
- Ophiocolea
- Oroxylum
- Pajanelia
- Pandorea
- Parabignonia
- Paragonia
- Parateconia
- Parmentiera
- Pauldopia
- Perianthomega
- Periarrabidaea
- Perichlaena
- Phryganocydia
- Phyllarthron
- Phyloctenium
- Piriadacus

Families of Dicotyledons

- Pithecoctenium
- Pleonotoma
- Podranea
- Potamoganos
- Pseudocatalpa
- Pyrostegia
- Radermachera
- Rhigozum
- Rhodocolea
- Roentgenia
- Romeroa
- Saritaea
- Sparattosperma
- Spathicalyx
- Spathodea
- Sphingiphila
- Spirotecoma
- Stereospermum
- Stizophyllum
- Tabebuia
- Tanaecium
- Tecoma
- Tecomanthe
- Tecomella
- Tourrettia
- Tynanthus
- Urbanolophium
- Xylophragma
- Zeyheria

Chapter – 121
Pedaliaceae

Classification (Bentham and Hooker)

Phanerogams
Dicotyledons
Gamopetalae
Bicarpellate
Personales
Pedaliaceae

General Characters

→ Mostly annual or perennial herbs, rarely shrubs
→ Branched Tap root
→ Leaves – opposite sometimes alternate, simple, entire or lobed, exstipulate, mucilaginous, glandular hairs present
→ Inflorescence – cymose, solitary axillary or simple axillary dichasial cyme (usually 3-flowered cymes). Rarely flowers – fascicled or in racemes.
→ Flower – Pedicellate, hermaphrodite (bisexual), zygomorphic, irregular, complete, hypogynous. Bracts may or may not be present
→ Calyx consists of 5 sepals (rarely 4), sepals connate at base, gamosepalous, aestivation –

imbricate
- → Corolla consists of 5 petals, somewhat bilabiate Corolla forms a broad tube, gamopetalous, aestivation imbricate
- → Androecium - 4 stamens, didynamous, sometimes the 5th posterior stamen is represented by a small staminode, epipetalous stamens, anthers - dithecous and dehisce by longitudinal slit, introrse. Sometimes stamen number is reduced to only 2.
- → Gynoecium - 2 carpels, syncarpous, 2 locules (but due to false septum formation, the locule number is doubled i.e. 4), ovary mostly superior, placentation axile, ovules one to many on each placenta, ovules anatropous. Style usually one, slender filiform. Stigma 2-lobed
- → Fruit - either a loculicidal capsule or nut, often spiny, winged, hooked or thorny.
- → Seeds albuminous with thin endosperm and a small straight embryo. Seeds smooth.
- → Pollination entomophilous

Genera included under Pedaliaceae

- Ceratotheca
- Dicerocaryum
- Harpagophytum
- Holubia
- Josephinia
- Linariopsis
- Pedaliodiscus
- Pedalium
- Pterodiscus
- Rogeria
- Sesamothamnus
- Sesamum
- Uncarina.

Chapter 122: Acanthaceae

Classification (Bentham and Hooker)

- Phanerogams
- Dicotyledons
- Gamopetalae
- Bicarpellatae
- Personales
- Acanthaceae

General Characters

→ Herbs or shrubs. Rarely trees. Some are xerophytic (*Barleria*, *Acanthus*). Some are climbing (*Mendoncia*). *Thunbergia alata* is a herbaceous twiner. *Acanthus elicifoleus* is a halophyte and represents mangrove vegetation.

→ Leaves: Simple, usually opposite decussate, entire, exstipulate, thin and delicate. In xerophytes, the leaf blade is more or less spiny.

→ The characteristic feature of this family is the presence of cystoliths in the epidermal cells of leaves and stems. Cystoliths are calcium carbonate crystals.

→ Inflorescence — cymose. Sometimes flowers are found in short axillary clusters. Frequently

Spikes and racemes are found. Bracts and bracteoles generally well developed and brightly coloured.

→ Flowers are bracteate, bracteolate, hermaphrodite, complete, zygomorphic (irregular) and hypogynous. Usually flowers are tetramerous or pentamerous.

→ Calyx - five, sometimes four or rarely 3 sepals. Usually gamosepalous. In _Thunbergia_, the sepals are reduced to a narrow beam and the bracteoles here serve the purpose of protection. Aestivation is either contorted or imbricate

→ Corolla consists of 5 rarely 4 petals, gamopetalous corolla often bilabiate (In _Ruellia_). In such cases, the upper lip of the corolla is erect and bifid at the apex and the lower lip is horizontal and 3 lobed forming the landing platform for the insects. Usually the corolla has a large or short slender corolla tube which passes above into an equally five lobed limb (_Thunbergia_). Inner side of the corolla lip possesses dense hairs. These hairs often extend to the mouth of the corolla. Aestivation is either contorted or imbricate

→ Androecium consists of 4 didynamous stamens. Rarely 2 stamens are found. Very rarely 5 (Pentstemonacanthus). In the case of 4 stamens, the 5th posterior stamen is reduced to a staminode or disappears completely. The filaments are generally quite free and project out from the mouth of the corolla tube. Stamens of Thunbergia are short and remain inside the corolla tube. Stamens are inserted in the corolla (epipetalous). Anthers two or one celled.

→ Gynoecium consists of 2 carpels, syncarpous. Ovary superior, bilocular, two to more ovules are found in each locules. Axile placentation, Style – long and slender and project out from the mouth of the corolla tube. Stigma – 2, small; Posterior stigma is generally reduced.

→ Fruits – bilocular capsule which dehisce loculicidally. Drupe (Mendoncia)

→ Seeds – small and many; funicle of the seeds form a papilla, the seeds are rounded. The funicle of the seeds form a hook like projection known as jaculator, in which the seed rests (Ruellia). These jaculators make the fruits burst and the seeds get dispersed in different directions. Pollination entamophilous.

Genera included under Acanthaceae

- Acanthopale
- Acanthopsis
- Acanthostelma
- Acanthura
- Acanthus
- Achyrocalyx
- Adhatoda
- Afrofittonia
- Ambongia
- Ancistranthus
- Ancistrostylis
- Andrographis
- Angkalanthus
- Anisacanthus
- Anisosepalum
- Anisostachya
- Anisotes
- Apassalus
- Aphanosperma
- Aphelandra
- Aphelandrella
- Ascotheca
- Asystasia
- Asystasiella
- Ballochia
- Barleria
- Barleriola
- Beleperone
- Benoicanthus
- Blechum
- Blepharis
- Borneacanthus
- Boutonia
- Brachystephanus
- Bravaisia
- Brillantaisia
- Buceragenia
- Calacanthus
- Calophanoides
- Calycacanthus
- Camarotea
- Carlowrightia
- Celerina
- Cephalacanthus
- Chaetacanthus
- Chalarothyrsus
- Chamaeranthemum
- Chlamydocardia
- Chlamydostachys
- Chroesthes
- Clinacanthus
- Clistax
- Codonacanthus
- Conocalyx
- Corymbostachy
- Cosmianthemum
- Crabbea
- Crossandra
- Crossandrella
- Cyclacanthus
- Cylindrosolenium
- Cyphacanthus
- Dactylostegium
- Danguya
- Dasytropis
- Dicharothece
- Dicladanthera
- Dicliptera
- Diclyplosandra
- Dipteracanthus

Families of Dicotyledons

- → Dischistocalyx
- → Dolichostachys
- → Drejera
- → Drejerella
- → Duosperma
- → Dyschoriste
- → Ecbolium
- → Echinacanthus
- → Encephalosphaera
- → Epiclastopelma
- → Eranthemum
- → Eremomastax
- → Eusiphon
- → Filetia
- → Fittonia
- → Forcipella
- → Forsythiopsis
- → Gastranthus
- → Geissomeria
- → Glossochilus
- → Golaea
- → Graphandra
- → Graptophyllum
- → Gymnophragma
- → Gymnostachyum
- → Gynocraterium
- → Gypsacanthus
- → Habracanthus
- → Hansteinia
- → Haplanthodes
- → Harpochilus
- → Henrya
- → Herpetacanthus
- → Heteradelphia
- → Holographis
- → Hoverdenia
- → Hulemacanthus
- → Hygrophila
- → Hypoestes
- → Ichthyostoma
- → Indoneesiella
- → Ionacanthus
- → Isoglossa
- → Isotheca
- → Jadunia
- → Juruasia
- → Justicia
- → Kalbreyeracanthus
- → Kalbreyeriella
- → Kosmosiphon
- → Kudoacanthus
- → Lankesteria
- → Lasiocladus
- → Leandriella
- → Lepidagathis
- → Leptostachya
- → Liberatia
- → Linariantha
- → Lindauea
- → Lophostachys
- → Louteridium
- → Lychniothyrsus
- → Mackaya
- → Marcania
- → Megalochlamys
- → Megalostoma
- → Megaskepasma
- → Melittacanthus
- → Mellera
- → Metarungia
- → Mexacanthus
- → Mimulopsis

→ Mirandea
→ Monothecium
→ Morsacanthus
→ Neohallia
→ Neriacanthus
→ Neuracanthus
→ Odontonema
→ Odontonemella
→ Ophiorrhiziphyllon
→ Oplonia
→ Oreacanthus
→ Orophochilus
→ Pachystachys
→ Pelecostemon
→ Penstemonacanthus
→ Perenideboles
→ Pericalypta
→ Periestes
→ Peristrophe
→ Petalidium
→ Phaulopsis
→ Phialacanthus
→ Phidiasia
→ Phlogacanthus
→ Physacanthus
→ Podorungia
→ Poikilacanthus
→ Polylychnis
→ Pranceacanthus
→ Pseuderanthemum
→ Pseudodicliptera
→ Pseudoruellia
→ Psilanthele
→ Ptyssiglotis
→ Pulchranthus
→ Pupilla
→ Razisea
→ Rhinacanthus
→ Rhombochlamys
→ Ritonia
→ Rostellularia
→ Ruellia
→ Ruelliopsis
→ Rungia
→ Ruspolia
→ Ruttya
→ Salpinctium
→ Salpixantha
→ Sebastiano-schaueria
→ Samuelsonia
→ Sanchezia
→ Santapaua
→ Sapphoa
→ Satanocrater
→ Sautiera
→ Schaueria
→ Schwabea
→ Sclerophyllum
→ Sclerochiton
→ Sephonoglossa
→ Spathacanthus
→ Sphacanthus
→ Sphinctacanthus
→ Spirostigma
→ Standleyacanthus
→ Steirosanchezia
→ Stenandriopsis
→ Stenandrium
→ Stenostephanus
→ Stephanophysum
→ Streblacanthus
→ Streptosiphon
→ Strobilanthes

- Strobilanthopsis
- Styasasia
- Suessenguthia
- Synchoriste
- Taeniandra
- Tarphochlamys
- Teliostachya
- Tessmanniacanthus
- Tetramerium
- Theileamea
- Thomandersia
- Thyanostigma
- Tremacanthus
- Triaenanthus
- Trechanthera
- Trichocalyx
- Ulleria
- Varala
- Vindasia
- Warpuria
- Xantheranthemum
- Xerothamnella
- Yeatesia
- Zygoruellia

Sub class: Gamopetalae

Series: Bicarpellatae

Order: Lamiales

Families:

Myoporineae

Selagineae

Verbenaceae

Labiatae

Plantagineae

Chapter - 123
Myoporineae

Classification (Bentham and Hooker)

Phanerogams
Dicotyledons
Gamopetalae
Bicarpellatae
Lamiales
Myoporineae

General Characters

→ Small trees or shrubs
→ Leaves - deciduous, alternate or opposite or whorled; spiral; often gland dotted; simple, epulvinate, lamina - entire; exstipulate Margin - entire / crenate / serrate / dentate
→ Plants hermaphrodite
→ Inflorescence - cymes
→ Flowers - ebracteate, regular to very irregular Pentamerous, tetracyclic.
→ Perianth with distinct calyx and corolla.
→ Calyx consists of 4-5 sepals; 1 whorled; Polysepalous, rarely gamosepalous (blunt lobed or toothed); regular; imbricate.
→ Corolla consists of 5 petals, 1 whorled,

gamopetalous, imbricate

→ Androecium 3-5 stamens, adnate to corolla tube, free of 1 another; 1 whorled; didynamous stamens; Anthers, introrse, versatile, dehiscing via longitudinal slits

→ Gynoecium – bicarpellary, syncarpous, superior; ovary 2 locular; Styles-1, apical; stigma-1, lobed; Placentation axile or apical; ovules 1-2 per locule; anatropous

→ Fruit – fleshy to non-fleshy, indehiscent, Schizocarp (drupe)

→ Seeds – scantily endospermic or non-endospermic

Genera included under Myoporaceae

- → Diocirea
- → Eremophila
- → Myoporum

Chapter-124
Selagineae

Classification (Bentham and Hooker)

    Phanerogams
    Dicotyledons
    Gamopetalae
    Bicarpellatae
    Lamiales
    Selagineae

General characters

→ Small shrubs or tufted perennial herbs rarely annuals
→ Leaves - Simple, alternate, whorled or rarely the lower opposite, cauline or rarely radical, Margins entire or toothed. Stipules absent
→ Inflorescence - terminal, elongated spikes, racemes or corymbose panicles
→ Flowers - zygomorphic, bisexual, Pentamerous.
→ Calyx fused and usually 5 lobed or 2-3 lobed or spathaceous
→ Corolla - tubular, usually 5 lobed, sometimes 2 lipped

→ Stamens 4, didynamous, epipetalous; anthers versatile, 1-locular, dehisce by longitudinal slits
→ Disc present, variable, ovary superior, 2-locular or by abortion 1 (rarely), style-terminal, filiform; ovary solitary
→ Fruit - indehiscent
→ 2-1 seeded nutlets.

Genera included under Selagineae
→ Hebenstretia
→ Selago.

Verbenaceae

Classification (Bentham and Hooker)

- Phanerogams
- Dicotyledons
- Gamopetalae
- Bicarpellatae
- Lamiales
- Verbenaceae

General Characters

→ Herbs, shrubs or trees. *Avicennia* is a mangrove shrub

→ Stem - quadrangular and possesses spines

→ Leaves - usually opposite or whorled, rarely alternate. Generally simple, very rarely compound. The leaves may be entire or divided and exstipulate

→ Inflorescence - great variation, racemose or cymose. Flowers may be arranged in racemes, spikes or head or may be axillary. Axillary cymes are aggregated in panicles and showy inflorescence.

→ Flower - Pentamerous, rarely showing reduction in calyx - tetramerous; hermaphrodite, rarely unisexual by means of abortion, zygomorphic, sometimes actinomorphic, complex, hypogynous.

→ Calyx consists of 5 sepals, gamosepalous, valvate aestivation; In the fruiting stage calyx becomes enlarged and much altered. Sometimes it become bladder like and encloses the fruit. Rarely sepals number range from 4 to 8.

→ Corolla consists of 5 petals. It is tubular often curved with a spreading limb, rarely campanulate, lobes equal, unequal or more or less two-lipped, aestivation imbricate

→ Androecium - 4 fertile stamens - didynamous, the stamens remain inserted on the corolla tube and alternate with its lobes; the 5th posterior is usually represented by a staminode or altogether absent. Filaments free, anthers dorsifixed, two celled, dehisce by longitudinal slits and introrse.

→ Gynoecium - 2 carpels, rarely 4 or 5 carpels. A false septum develops from the middle of each carpel inwards dividing the two carpels into four one ovuled chambers. In the young, placentation is parietal but later on it becomes axile.

→ Fruit - drupe.

→ Seeds are exalbuminous, ovate and sessile. Seed coat is thin and membranous. Embryo is straight and the radicle small directed towards the micropyle

→ Pollination entomophilous.

Genera Included under Verbenaceae

- Acantholippia
- Adelosa
- Aegiphila
- Aloysia
- Amasonia
- Archboldia
- Asepalum
- Baillonia
- Bouchea
- Burroughsia
- Callicarpa
- Caryopteris
- Casselia
- Chascanum
- Citharexylum
- Clerodendrum
- Coelocarpum
- Cornutia
- Dimetra
- Diostea
- Dipyrena
- Duranta
- Faradaya
- Garrettia
- Geunsia
- Glandularia
- Glossocarya
- Gmelina
- Hierobotana
- Holmskioldia
- Hosea
- Huxleya
- Hymenopyramis
- Junellia
- Karomia
- Lampaya
- Lantana
- Lippia
- Monochilus
- Nashia
- Neorapinia
- Neosparton
- Nesogenes
- Oncinocalyx
- Oxera
- Paravitex
- Parodianthus
- Peronema
- Petitia
- Xeroaloysia
- Petraeovitex
- Petraea
- Phyla
- Pitraea
- Premna
- Priva
- Pseudocarpidium
- Recordia
- Rehdera
- Rhaphithamnus
- Schnabelia
- Stachytarpheta
- Styloden
- Surfacea
- Tamonea
- Tectona
- Teijsmanniodendron
- Tetraclea
- Teucridium
- Tsoongia
- Ubochea
- Urbania
- Verbena
- Verbenoxylum
- Vitex
- Viticipremna
- Xolocotzia

Chapter 126: Lamiaceae

Classification (Bentham and Hooker)

- Phanerogams
- Dicotyledons
- Gamopetalae
- Bicarpellatae
- Lamiales
- Lamiaceae (Labiatae)

General characters

→ Annual or perennial herbs. Certain plants are xerophytes with extremely reduced leaves. Rarely trees; rarely climbers. In *Rosmarinus*, the leaves are rolled back and stomata are found among all hairs in the grooves on the underside of the leaf.

→ Stem – young shoots are usually four sided or quadrangular.

→ Leaves – simple, opposite, decussate and exstipulate. Many variations from entire blade to toothed, lobed, cut or finely dissected. In *Ocimum*, whorled leaf arrangement is present. All parts of the plant are more or less hairy and possess glandular hairs, which secrete characteristic scent of the genus. Sessile scented oil secreting glands are also found frequently on the epidermis

→ Inflorescence (characteristic) – Verticillaster. In this type the whorls of flowers develop at the nodes. This consists of a pair of cymose inflorescences. Each cyme forms a simple 3 flowered dichasium. (*Salvia*). Sometimes branching takes place and pair of monochasial cyme develops (*Lamium*). In other cases the main axis or lateral axis are more or less developed. In some, all axes are reduced developing a dense sessile inflorescence. (*Leucas*)

→ Flowers – hermaphrodite, zygomorphic, rarely actinomorphic, hypogynous

→ Calyx consists of 5 sepals, gamosepalous; sepals inferior, persistent, campanulate or tubular, with free teeth or lobes, sometimes bilabiate (*Thymus*). In *Hoslundia*, the persistent sepals become fleshy in the fruit. Aestivation – valvate, imbricate or rarely quincuncial.

→ Corolla consists of 5 petals, gamopetalous, tubular and limb variously bilabiate.

Corolla consists of 2 pairs viz. tube and limb. The tube is straight or bent and wide towards the mouth. The limb is rarely equally fine toothed. Aestivation – Imbricate.

→ Androecium – Stamens 4, didynamous, sometimes reduced to two. They are epipetalous and alternate with the corolla lobes; 5th posterior stamen is usually absent and rarely developed. Usually the anterior pair of the stamen is longer than the posterior pair. The anthers two-celled; dehisce by longitudinal slits. In between the two cells of the anther, connective is present. A four lobed hypogynous disc is present. Anterior lobes of the disc secrete nectar.

→ Gynoecium consists of 2 median carpels, syncarpous, seated on a hypogynous nectar secreting disc; ovary-bilocular but later a constriction develops which divides the two carpels into 4; Ovules – anatropous. Style arises from the base of the ovary in between the four loculi and is known as gynobasic style. Stigmatic papillae of bilobed stigma are situated at the tip of the style arms; Placentation axile; Ovary superior.

→ Fruit - consists of 4 one seeded nutlets, included within the persistent calyx
→ Seeds - exalbuminous or with scanty endosperm
→ Pollination - entomophilous.

Genera Included under Lamiaceae

- Acanthomintha
- Achyrospermum
- Acinos
- Acrocephalus
- Acrotome
- Acrymia
- Aeollanthus
- Agastache
- Ajuga
- Ajugoides
- Alajja
- Alvesia
- Amethystea
- Anisochilus
- Anisomeles
- Antonina
- Asterohyptis
- Ballota
- Basilicum
- Becium
- Benguellia
- Blephilia
- Bostrychanthera
- Bovonia
- Brazoria
- Bystropogon
- Calamintha
- Capitanopsis
- Capitanya
- Catoferia
- Cedronella
- Ceratanthus
- Chamaesphacos
- Chaunostoma
- Chelonopsis
- Cleonia
- Clinopodium
- Colebrookia
- Coleus
- Collinsonia
- Colquhounia
- Comanthosphace
- Conradina
- Craniotome
- Cuminia
- Cunila
- Cyclotrichium
- Cymaria
- Dauphinea
- Dicerandra
- Dorystaechas
- Dracocephalum
- Drepanocaryum
- Eichlerago
- Elsholtzia
- Endostemon
- Englerastrum
- Eremostachys
- Eriope
- Eriophyton
- Eriopidion
- Eriothymus
- Erythrochlamys
- Eurysolen
- Fuerstia
- Galeopsis
- Genrosporum
- Glechoma
- Glechon
- Gomphostemma

- → Gontscharovia
- → Hanceola
- → Haplostachys
- → Haumaniastrum
- → Hedeoma
- → Hemiandra
- → Hemigenia
- → Hemizygia
- → Hesperozygis
- → Heterolamium
- → Hoehnea
- → Holocheila
- → Holostylon
- → Horminum
- → Hoslundia
- → Hymenocrater
- → Hypenia
- → Hypogomphia
- → Hyptidendron
- → Hyptis
- → Hyssopus
- → Isodictyophorus
- → Isodon
- → Isoleucas
- → Keiskea
- → Kinostemon
- → Kudrjaschevia
- → Kurzamra
- → Lagochilus
- → Lagopsis
- → Lallemantia
- → Lamiophlomis
- → Lamium
- → Lavandula
- → Leocus
- → Leonotis
- → Leonurus
- → Lepechinia
- → Leucas
- → Leucosceptrum
- → Limniboza
- → Lophanthus
- → Loxocalyx
- → Lycopus
- → Macbridea
- → Marmoritis
- → Marrubium
- → Marsypianthes
- → Meehania
- → Melissa
- → Melittis
- → Mentha
- → Meriandra
- → Mesona
- → Metastachydium
- → Microcorys
- → Micromeria
- → Microtoena
- → Minthostachys
- → Moluccella
- → Monarda
- → Monardella
- → Mosla
- → Neoeplingia
- → Neohyptis
- → Nepeta
- → Neustruevia
- → Nosema
- → Notochaete
- → Ocimum
- → Octomeron
- → Ombrocharis

- Origanum
- Orthosiphon
- Otostegia
- Panzerina
- Paraeremostachys
- Paralamium
- Paraphlomis
- Peltodon
- Pentapleura
- Perilla
- Perilomia
- Perovskia
- Perrierastrum
- Phlomis
- Phlomoides
- Phyllostegia
- Physostegia
- Piloblephis
- Piloblephis
- Pitradia
- Platostoma
- Plectranthus
- Pogogyne
- Pogostemon
- Poliomintha
- Pozophora
- Prasium
- Prostanthera
- Prunella
- Pseuderemostachys
- Puntia
- Pycnanthemum
- Pycnostachys
- Renschia
- Rhabdocaulon
- Raphidion
- Rhododon
- Rosmarinus
- Rostrinucula
- Roylea
- Rubiteucris
- Sabaudia
- Saccocalyx
- Salvia
- Satureja
- Schizonepeta
- Scutellaria
- Sideritis
- Siphocranion
- Zhumeria
- Skapanthus
- Solenostemon
- Stachyopsis
- Stachys
- Stenogyne
- Sulaimania
- Suzukia
- Symphostemon
- Synandra
- Syncolostemon
- Tetradenia
- Teucrium
- Thorncroftia
- Thuspeinanta
- Thymbra
- Thymus
- Tinnea
- Trichostema
- Wenchengia
- Westringia
- Wiedemannia
- Wrixonia
- Zataria

Chapter - 127
Plantagineae

Classification (Bentham and Hooker)

Phanerogams
Dicotyledons
Gamopetalae
Bicarpellatae
Lamiales
Plantagineae

General characters

→ Herbs, shrubs, few aquatic plants with roots.
→ Leaves - spiral to opposite and simple to compound.
→ Tetramerous (4 sepals & 4 petals) (or) 5-8 merous; Flowers are polysymmetric. Corolla is often two-lipped. In some genera, androecium is formed before the corolla.
→ Fruit - capsule which dehisces through the partitions between the cells.

Genera included under plantagineae

→ Angelonia
→ Basistemon
→ Melosperma
→ Monopera
→ Monttea
→ Ourisia

- Acanthorrhinum
- Albraunia
- Anarrhinum
- Antirrhinum
- Asarina
- Chaenorhinum
- Cymbalaria
- Epixiphium
- Galvezia
- Gambelia
- Holmgrenanthe
- Holzneria
- Howelliella
- Kickxia
- Linaria
- Lophospermum
- Mabrya
- Maurandya
- Misopates
- Mohavea
- Neogaerrhinum
- Nuttallanthus
- Pseudomisopates
- Pseudorontium
- Rhodochiton
- Sairocarpus
- Schweinfurthia
- Callitriche
- Hippuris
- Brookea
- Chelone
- Chionophila
- Collinsia
- Keckiella
- Nothochelone
- Penstemon
- Tonella
- Uroskinnera
- Digitalis
- Erinus
- Campylanthus
- Globularia
- Poskea
- Achetaria
- Adenosma
- Bacopa
- Benjaminia
- Boelckea
- Capraria
- Cheilophyllum
- Conobea
- Darcya
- Deinostema
- Dizygostemon
- Dopatrium
- Fonkia
- Geochorda
- Gratiola
- Hydranthelium
- Hydrotriche
- Ildefonsia
- Leucospora
- Limnophila
- Maeviella
- Mecardonia
- Otacanthus
- Philcoxia
- Schistophragma
- Schizosepala
- Scoparia
- Sophronanthe
- Stemodia
- Tetraulacium
- Hemiphragma
- Plantago
- Aragoa

- → Littorella
- → Russelia
- → Tetranema
- → Ellisiophyllum
- → Sibthorpia
- → Chionohebe
- → Detzneria
- → Hebe
- → Kashmiria
- → Lagotis
- → Neopicrorhiza
- → Paederota
- → Parahebe
- → Picrorhiza
- → Scrofella
- → Synthyris
- → Veronica
- → Veronicastrum
- → Wulfenia
- → Wulfeniopsis.

Sub class: Monochlamydeae

Series: Curvembryae

Families:

Nyctagineae

Illecebraceae

Amaranthaceae

Chenopodiaceae

Phytolaccaceae

Batideae

Polygonaceae

Chapter – 128
Nyctaginaceae

Classification (Bentham and Hooker)

- Phanerogams
- Dicotyledons
- Monochlamydeae
- Curvembryae
- Nyctagineae

General Characters:

→ Herbs, shrubs or trees. Some climbers (Bougainvillaea)

→ Needle like raphides of calcium oxalate are found in various organs of the plant. Anamolous secondary growth is present

→ Leaves - usually opposite, simple, entire, exstipulate. Opposite leaves of a pair are usually very unequal in size.

→ Inflorescence - cymose type. In Bougainvillaea, there are three large, coloured petaloid bracts which surround a group of three flowers. In some plants (Boerhaavia), the involucral bracts are reduced to teeth or scale like structures.

→ Flowers :- bisexual mostly but in Pisonia by suppression of stamens or pistils they become unisexual; regular (actinomorphic) and hypogynous

→ Perianth consists of five petaloid perianth leaves.

It is tubular and remains persistent in the fruit. The persistent perianth which envelops the fruit becomes leathery, hardened or fleshy. The mucilage or glandular hairs develop on this portion and serve as a means of dissemination (Pisonia).

→ Androecium consists of 5 stamens that alternate the perianth leaves. The number of stamens range from 1 to 30. The filaments are of unequal length.

→ Gynoecium consists of single carpel. Ovary - free, superior, unilocular. It bears a long simple style and a single basal erect ovule. The ovule may be anatropous and campylotropous.

→ Fruit is an achene which remains enclosed within the persistent perianth.

→ Seeds are endospermic and contains a large, straight, curved or folded embryo. The cotyledons are unequal in size.

→ Pollination is entamophilous.

Genera included under Nyctaginaceae

- Abronia
- Acleisanthes
- Allionia
- Ammocodon
- Andradea
- Anulocaulis
- Belemia
- Boerhavia
- Bougainvillea
- Caribea
- Cephalatomandra
- Colignonia
- Commicarpus
- Cryptocarpus
- Cuscatlania
- Cyphomeris
- Guapira
- Grajalesia
- Izabalaea
- Leucaster
- Mirabilis
- Neea
- Neeopsis
- Nyctaginia
- Okenia
- Phaeoptilum
- Pisonia
- Pisonella
- Ramisia
- Reichenbachia
- Salpianthus
- Selinocarpus
- Tripterocalyx

Chapter-129
Illecebraceae

Classification (Bentham and Hooker)
- Phanerogams
- Dicotyledons
- Monochlamydeae
- Curvembryae
- Illecebraceae

General characters
→ Herbs or subshrubs
→ Leaves - opposite, often joined at base; simple, entire; stipules scarious or absent
→ Inflorescence - cymose, sometimes raceme like or densely congested or flowers solitary
→ Flowers - regular, 4-5-8-merous, receptacle sometimes produced into a cupular perigonium.
→ Sepals free or joined into calyx tube
→ Petals as many as sepals, free, sometimes minute or absent
→ Stamens usually twice as many as sepals in 2 whorls, innermost often represented by staminodes

anthers 2 celled, dehiscing longitudinally.
→ Ovary superior, sometimes on stalk (carpophore), 1 celled or incompletely septate at base, placentation free central or basal, ovules 1-many, styles 2-5, free or connate
→ Fruit - Thin walled 1-seeded Utricle
→ Seeds often cochleate or reniform with a curved embryo, often tuberculate or papillate

Genera included under Illecebraceae

→ Corrigiola
→ Telephium
→ Herniaria
→ Paronychia
→ Achyronychia
→ Cardionema
→ Chaetonychia
→ Cometes
→ Orchenanthus
→ Gymnocarpos
→ Illecebrum
→ Lochia
→ Phelippiella
→ Pollichia
→ Pteranthus
→ Sclerocephalus
→ Scopulophila
→ Sphaerocoma
→ Loeflingia
→ Polycarpaea
→ Polycarpon
→ Spergula
→ Spergularia

Chapter - 130
Amaranthaceae

Classification (Bentham and Hooker)

- Phanerogams
- Dicotyledons
- Monochlamydeae
- Curvembryae
- Amaranthaceae

General Characters

→ Mostly annual or perennial herbs or shrubs.

→ Leaves - opposite or alternate, simple, entire, exstipulate, usually covered with hairs.

→ Inflorescence - usually racemose type. They may be simple or branched spike or raceme. Mostly the small flowers are arranged in dense fascicles.

→ Flowers - very minute; each flower bears a pair of large membranous persistent bracteoles, which are sterile; bracteate, bracteolate, hermaphrodite, rarely unisexual by abortion, actinomorphic and hypogynous.

→ Perianth - dry, membranous and very often white or coloured. It consists of 4 or 5 perianth leaves which are green and herbaceous. Tepals are free or more or less united. Sometimes they become hairy.

→ Androecium consists of 4 or 5 stamens, situated opposite to tepals. Usually the stamens are united at the base forming a membranous tube. Sometimes these stamens bear fringed outgrowths in between them. Mostly anthers two-celled, rarely one celled. Dehisce longitudinally.

→ Gynoecium consists of 2-3 carpels, syncarpous. Ovary- unilocular and superior. In *celosia* and other genera there is a single basal ovule. Ovule- camphylotropous; Styles- 1, 2 or 3.

→ Fruit- dry. They may be nut, drupe or berry.

→ Seeds- endospermic with rough or polished testa. Embryo is curved and lies close to the seed coat.

→ Pollination entomophilous.

Genera included under Amaranthaceae

- Achyranthes
- Achyropsis
- Acnida
- Aerva
- Allmania
- Alternanthera
- Amaranthus
- Arthraerua
- Bosea
- Brayulinea
- Calicorema
- Celosia
- Centema
- Centemopsis
- Centrostachys
- Charlessoa
- Charpentiera
- Chionothrix
- Cyathula
- Dasysphaera
- Deeringia
- Digera
- Eriostylos
- Froelichia
- Gomphrena
- Gossypianthus
- Guilleminea
- Hebanthe
- Hemichroa
- Henonia
- Herbstia
- Hermbstaedtia
- Indobanalia
- Irenella
- Iresine
- Kyphocarpa
- Lagrezia
- Leucosphaera
- Lithophila
- Lopriorea
- Marcelliopsis
- Mechowia
- Nelsia
- Neocentema
- Nothosaerva
- Nothotrichium
- Nyssanthes
- Pandiaka
- Pfaffia
- Philoxerus
- Pleuropetalum
- Pleuropterantha
- Polyrhabda
- Pseudogomphrena
- Pseudoplantago
- Pseudosericocoma
- Psilotrichopsis
- Psilotrichum
- Ptilotus
- Pupalia
- Quaternella
- Rosifax
- Saltia
- Sericocoma
- Sericocomopsis
- Sericorema
- Sericostachys
- Spamosia
- Stilbanthus
- Tidestromia
- Trichuriella
- Volkensinia
- Woehleria
- Xerosiphon

Chapter-131
Chenopodiaceae

Classification (Bentham and Hooker)

Phanerogams
Dicotyledons
Monochlamydeae
Curvembryae
Chenopodiaceae

General characters:

→ Plants are annual or biennial herbs and rarely shrubs are also present. Small trees are very rare (*Haloxylon*). *Salsola Kali* is known as salt wort and grows near sea shores. Some grow in marshy places (*Salicornia*).

→ Branched tap root system.

→ Stem - erect, herbaceous, branched, cylindrical, solid, hairy and green.

→ Leaves - simple and alternate, rarely opposite. In many cases leaves are minute, fleshy and covered with hairs.

→ Inflorescence - cymose, dichasial cyme.

→ Flowers - Minute, greenish in colour and arranged in dense cymose inflorescence. Hermaphrodite or unisexual. Plants may be monoecious or dioecious; bracteate, actinomorphic

and hypogynous.
- → Perianth consists of 5 perianth leaves (tepals), sepaloid, rarely 3 or 4 tepals. Tepals may be free or united. Usually the perianth leaves remain persistent in the fruit. Sometimes male flowers lack tepals. Aestivation – imbricate.
- → Androecium consists of 5 stamens, usually the stamen number corresponds to that of tepals. The stamens are hypogynous or situated on a disc. Anthers – bi celled and dehisce longitudinally.
- → Gynoecium – 2 or 3 carpels, syncarpous. Ovary – superior, rarely inferior or half inferior. It is unilocular with a single basal campylotropous ovule.
- → Fruit – nut or achene enclosed in the perianth.
- → Seeds – albuminous with a curved or twisted embryo.
- → Pollination is entomophilous.

Genera included under Chenopodiaceae

- Acroglochin
- Aellenia
- Agathophora
- Agriophyllum
- Alexandra
- Allenrolfea
- Anabasis
- Anthochlamys
- Aphanisma
- Archiatriplex
- Arthrocnemum
- Arthrophytum
- Atriplex
- Axyris
- Baolia
- Bassia
- Beta
- Bienertia
- Blitum
- Borsczowia
- Camphorosma
- Ceratocarpus
- Chenopodium
- Climacoptera
- Corispermum
- Cornulaca
- Cremnophyton
- Cyathobasis
- Cycloloma
- Didymanthus
- Dissocarpus
- Einadia
- Enchylaena
- Eremiophea
- Eriochiton
- Exomis
- Fadenia
- Fredolia
- Gamanthus
- Girgensohnia
- Grayia
- Habletzea
- Halanthium
- Halarchon
- Halimione
- Halimocnemis
- Halocharis
- Halocnemum
- Halogeton
- Halopeplis
- Halosarcia
- Halostachys
- Halothamnus
- Haloxylon
- Hammada
- Hemichroa
- Heterostachys
- Holmbergia
- Horaninovia
- Iljinia
- Kalidium
- Kirilowia
- Kochia
- Krascheninnikovia
- Lagenantha
- Maireana
- Malacocera
- Manochlamys
- Microcnemum
- Microgynoecium
- Monolepis
- Nanophyton
- Neobassia
- Nitrophila

- Noaea
- Nucularia
- Ofaiston
- Oreobleton
- Osteocarpum
- Pachycornia
- Panderia
- Petrosimonia
- Physandra
- Piptoptera
- Polycnemum
- Rhagodia
- Raphidophyton
- Roycea
- Salicornia
- Salsola
- Sarcocornia
- Sclerobletum
- Sclerochlamys
- Sclerolaena
- Sclerostegia
- Seedletzia
- Sevada
- Spinacia
- Stelligera
- Suaeda
- Suckleya
- Sympegma
- Tecticornia
- Teloxys
- Threlkedia
- Traganopsis
- Traganum
- Zuckia

Chapter-132
Phytolaccaceae

Classification (Bentham and Hooker)

- Phanerogams
- Dicotyledons
- Monochlamydeae
- Curvembryae
- Phytolaccaceae

General characters

→ Usually herbs; some trees, shrubs or leanas. More or less succulent or non-succulent

→ Leaves - alternate, spiral; herbaceous or fleshy; petiolate or sessile; simple; entire; usually exstipulate rarely stipulate; stipules spiny; Margin - entire

→ Plants hermaphrodite or dioecious

→ Flowers solitary or in inflorescence - racemes or spikes or cymes or panicles.

→ Flowers - bracteate, bracteolate, small, regular, cyclic. Floral receptacle developing on a gynophore. Hypogynous disk present in some.

→ Perianth sepaline. Calyx - 4 or 5-10 sepals, 1 whorled, usually persistent, imbricate

→ Androecium - 4 - Many stamens; free of one another or basally connate; stamens isomerous with perianth or diplostemonous; Anthers - dorsifixed, introrse, dehiscing via longitudinal slits.

→ Gynoecium - 1 to 16 carpelled, syncarpous, superior 1 ovuled; placentation basal; ovules 1 per locule.

→ Fruit - fleshy or non-fleshy; aggregate or non-aggregate; indehiscent; samaroid or nucular or drupaceous; dehiscent or indehiscent or schizocarp, capsule or berry.

→ Seeds - with starch.

Genera included under Phytolaccaceae

- Anisomeria
- Ercilla
- Gallesia
- Hilleria
- Ledenbergia
- Lophiocarpus
- Microtea
- Monococcus
- Petiveria
- Phytolacca
- Rivina
- Schindleria
- Seguieria
- Trichostigma.

Chapter-133
Batideae

Classification (Bentham and Hooker)

- Phanerogams
- Dicotyledons
- Monochlamydeae
- Curvembryae
- Batideae

General characters

→ Weak, shrubs; Plants succulent; xerophytic

→ Leaves - opposite, flat or terete, fleshy; subsessile to sessile; slightly connate; simple; lamina - entire, linear or oblong o... obovate. Leaves - exstipulate or stipulate

→ Stipules represented by glands.

→ Plants monoecious or dioecious.

→ Inflorescence - spikes.

→ Flowers - bracteate, somewhat irregular. Tetramerous; male flowers cyclic

→ Perianth with distinct calyx and corolla (male flowers) and absent (female flowers); 1 or 2 whorled. gamosepalous, blunt-lobed, cupuliform or campanulate; bilabiate

→ Corolla of male flowers - 4 petals; 1 whorled

Polypetalous or gamopetalous, regular; petals clawed.
→ Androecium – 4 stamens, free of one another, 1 whorled; isomerous with perianth; Anthers – dorsifixed, versatile, dehiscing via a longitudinal slits, introrse
→ Gynoecium – bicarpellary, syncarpous, superior; ovary sessile; stigma – 2; placentation basal; ovules – 1 per locule; funicled; anatropous.
→ Fruit – fleshy, indehiscent, drupe.
→ Seeds – non-endospermic

Genera included under Batideae
→ Batis

Chapter-134
Polygonaceae

Classification (Bentham and Hooker)

Phanerogams
Dicotyledons
Monochlamydeae
Curvembryae
Polygonaceae

General Characters

→ Mostly annual or perennial herbs. Climbing species also occur. Xerophytes (*Polygonum*); aquatic.

→ Stem - Mostly herbaceous, sometimes woody. It is very often conspicuously swollen at nodes. *Muehlenbeckia platyclada*, the stem and its branches are flattened forming flat phylloclades or cladodes which are jointed at the nodes.

→ Leaves - Scattered, mostly alternate, rarely opposite, simple and entire. Sometimes they are lobed (*Rheum*). Most characteristic feature of this family is the constant presence of an ochrea (sheathing stipule). Calcium oxalate crystals are frequently found in the cells.

→ Inflorescence - Racemose type. Sometimes solitary. *Eriogonum* - flowers in cymose umbels or heads.

Polygonum – Solitary or Panicled racemes. **Fagopyrum** – Capitate cymes.

→ Flowers are generally bisexual, rarely unisexual, actinomorphic (regular), small, trimerous. More rarely dimerous, cyclic and acyclic, usually borne in large number in compound inflorescences. Cyclic – P 3+3, A 3+3 G($\underline{3}$). Very rarely the cyclic flowers are dimerous (*Oxyria*), the outer staminal whorl is doubled. Outer stamens are branched in some (*Rheum*). Acyclic arrangement → P 5, A 5-8, G($\underline{3}$).

→ Perianth - Flowers with cyclic arrangement possess 6 similar perianth leaves arranged in 2 whorls of 3 each. In flowers of acyclic arrangement perianth leaves are arranged according to 2/5 phyllotaxy. Perianth leaves are alike, green, white or red. They may be free or more or less connate. In *Triplaris*, the 3 outer perianth leaves form long, flat, membranous, erect wings which look like a shuttlecock. This structure helps in the dissemination of the fruit.

→ Androecium - the number of stamens range from 6 to 9, sometimes fewer. Usually stamens are arranged in 2 whorls. The outer whorl consists of six stamens (antrorse) and the inner of ♂ stamens. In certain species of polygonum, the outgrowths of the receptacle, which are like the glands are also present within the whorl of stamens. Filaments are either free or connate at the base. Anthers dithecous and dehisce by longitudinal slits.

→ Gynoecium – tricarpellary rarely of 2 or 4 carpels; syncarpous. Ovary superior, unilocular and has a single erect orthotropous ovule. It is sessile, compressed and triangular. Simple style with 2-4 stigmas, according to the number of carpels. At the base of the ovary there is a nectar secreting annular disc. Placentation basal.

→ Fruit – Generally 3 sided, but sometimes 2 sided, dry, one seeded nut. <u>Coccoloba</u> bears a fleshy perianth.

→ Seeds – endospermic, curved or straight embryo remains more or less embedded in the endosperm.

→ Pollination is anemophilous or entomophilous.

Families of Dicotyledons

Genera included under Polygonaceae.

- Afrobrunnichia
- Antigonon
- Aristocapsa
- Atraphaxis
- Brunnichia
- Calligonum
- Centrostegia
- Chorizanthe
- Coccoloba
- Dedeckera
- Dodecahema
- Emex
- Eriogonum
- Fagopyrum
- Fallopia
- Gilmania
- Goodmania
- Gymnopodium
- Harfordia
- Hollisteria
- Knorringia
- Koenigia
- Lastarriaea
- Leptogonum
- Muconea
- Muehlenbeckia
- Nemacaulis
- Neomillspaughia
- Oxygonum
- Oxyria
- Oxytheca
- Parapteropyrum
- Persicaria
- Podopterus
- Polygonella
- Polygonum
- Pteropyrum
- Pterostegia
- Rheum
- Rumex
- Ruprechtia
- Stenogonum
- Symmeria
- Systenotheca
- Triplaris.

Sub class: Monochlamydeae
Series: Multivulatae Aquaticae
Families:

Podostemaceae

Chapter - 135
Podostemaceae

Classification (Bentham and Hooker)

Phanerogams
Dicotyledons
Monochlamydeae
Multiovulatae Aquaticae
Podostemataceae

General Characters

→ Species are found in streams of running water where they are attached to rocks, by their Moss like or lichen. Roots of the plants often photosynthetic and creeping.

→ Leaves - alternate, usually simple, linear to lamellate, often basally sheathing. Leaves small and of various form ranging from scaly to moss like. Sometimes leaves large, simple and dissected.

→ Inflorescence - Solitary and terminal, cymose

→ Flowers - Small, bisexual, actinomorphic, sometimes zygomorphic, haplochlamydous (i.e. Perianth in single whorl) or achlamydous (i.e. lack perianth) and hypogynous; flowers are usually produced only when plants are exposed by low water; fragrant, pedicellate.

→ Perianth is not differentiated into calyx and corolla. 3-5 perianth members or tepals which may be either free or sometimes somewhat united at the base, imbricate or valvate; sometimes tepals very small, scale like and numerous. In achlamydous flowers the young flower buds remain covered by thin membranes known as 'spathella'.

→ Androecium - Stamens 1-many, in one to several series, free; sometimes filaments of stamens are united into a tubular sheath round the pistil, anthers typically 4 celled, introrse, one staminode on each side of the stamens, dehiscence longitudinal lateral.

→ Gynoecium - Carpels 2-3, syncarpous, united into a bilocular or trilocular superior ovary; Placentation axile, sometimes 1-loculed ovary with free central placentation; the ovules are numerous and anatropous, styles 2-3 filiform or very short, often strongly papillate, usually distinct, styles as many as carpels; stigmas as many as styles, fimbriate or capitate

→ Fruit - Septicidal capsule
→ Seeds - Numerous, minute, non-endospermic

Genera included under Podostemaceae

- Angolaea
- Apinagia
- Buturia
- Castelnavia
- Cladopus
- Crenias
- Dalzellia
- Devillea
- Dicraeanthus
- Diplobryum
- Djinga
- Endocaulos
- Farmeria
- Hydrobryopsis
- Hydrobryum
- Indotristicha
- Jenmaniella
- Lawia
- Ledermanniella
- Lepothylax
- Letestuella
- Lonchostephus
- Lophogyne
- Macrarenia
- Macropodiella
- Malaccotristichia
- Marathrum
- Monostylis
- Mourera
- Oserya
- Paleodicraeia
- Podostemum
- Polypleurella
- Polypleurum
- Rhyncholacis
- Saxicolella
- Sphaerothylax
- Stonesia
- Thelethylax
- Torrenticola
- Tristicha
- Tulasneantha
- Weddellina
- Willisia
- Winklerella
- Zehnderia
- Zeylandium

Sub class: Monochlamydeae

Series: Multivulatae terrestres

Families:

Nepenthaceae

Cytinaceae

Aristolochieae

Chapter - 136
Nepenthaceae

Classification (Bentham and Hooker)

Phanerogams
Dicotyledons
Monochlamydeae
Multiovulatae Terrestres
Nepenthaceae

General characters

→ Shrubs or leanas or herbs. Plants carnivorous; Traps consists of 'pitchers'; Epiphytic or climbing. Heterophyllous.; Mostly tendril climbers.

→ Leaves - alternate, spiral; Petiolate (petiole winged); sheathing; simple; entire; exstipulate

→ Plants dioecious. Floral nectaries present; Pollination entomophilous.

→ Inflorescence - cymes, racemes and Panicles.

→ Flowers - ebracteate, ebracteolate; small, fragrant, regular, cyclic

→ Perianth - sepaline; 3-4 free; 2 whorled; usually isomerous; calyx - 3-4 sepals; 2 whorled, usually polysepalous, regular, imbricate

→ Androecium → 2-25 stamens; free of the Periant; Monadelphous.

Anthers – extrorse; dehiscing via longitudinal slits.

→ Gynoecium 3-4 carpelled, syncarpous, superior; ovary 3-4 locular; style-1 (short); stigma-1 (capitate to subpeltate). Placentation axile; ovules many per locule; anatropous

→ Fruit – non fleshy (leathery), dehiscent, elongated capsule (loculicidal)

→ Seeds endospermic

Genera included under Nepenthaceae

→ Nepenthes   → Anurosperma

Chapter-137
Cytinaceae

Classification (Bentham and Hooker)

- Phanerogams
- Dicotyledons
- Monochlamydeae
- Multiovulatae Terrestres
- Cytinaceae

General characters

→ Very peculiar endoparasitic herbs; vegetative body filamentous or fungoid; leaves - much reduced or absent; plants parasitic, rootless; endoparasitic.

→ Leaves - alternate or opposite or whorled; membranous (scales)

→ Plants monoecious or dioecious; Pollination - entomophilous

→ Flowers in inflorescence - spikate

→ Flowers - regular, cyclic; floral receptacle developing on an androphore or gynophore or not

→ Perianth - sepaline or petaline or of tepals; 4, joined forming a tube, 1 whorled, somewhat fleshy.

→ Androecium 15-many; united with gynoecium or not; free of 1 another; 1 whorled; sessile anthers; dehisce via longitudinal slits; extrose; unilocular.

→ Gynoecium 4-8 carpelled, syncarpous, inferior; ovary 4-8 locular; style-1, apical; stigmas truncate; placentation parietal; ovules many per locule.

→ Fruit- fleshy, indehiscent, berry.

→ Seeds- non-endospermic

Genera included under cytinaceae
→ Bolallophytum → cytinus

Chapter – 138
Aristolochieae

Classification (Bentham and Hooker)

Phanerogams
Dicotyledons
Monochlamydeae
Multiovulatae Terrestres
Aristolochieae

General characters

→ Shrubs or leanas or herbs; bearing essential oils; perennial; climbing or self supporting; mostly stem twiners.

→ Leaves – alternate, spiral, flat, herbaceous or membranous, petiolate, sheathing to non-sheathing; aromatic; simple; lamina usually entire rarely dissected; exstipulate.

→ Plants hermaphrodite. Pollination – entomophilous

→ Flowers solitary or inflorescence – cymes or racemes or spikes.

→ Flowers – small to large; often malodorous or odourless; regular to very irregular; tricyclic to pentacyclic

→ Perianth with distinct calyx and corolla or petaloid

→ Calyx consists of 3 or 6 sepals, 1 or 2 whorled;

→ Calyx - 3 sepals, 1 whorled, gamosepalous; entire or blunt-lobed; campanulate or tubular. Corolla when present, 3 petals, 1 whorled.

→ Androecium 6-36 stamens; united with the gynoecium. or free; free of one another or joined (monadelphous); 1 or 2 whorled. Anthers - filantherous or sessile anthers. Anthers cohering or separate from one another; basifixed or adnate; non-versatile; dehisce via longitudinal slits; extrorse

→ Gynoecium 4-6 carpelled; syncarpous; partly inferior or inferior. Epigynous. disk present or absent; styles 1 or 4-6 free or partly joined, apical; stigma - dry type. placentation - parietal, axile. Many ovules per locule

→ Fruit - non fleshy or rarely fleshy, dehiscent or rarely indehiscent; or a schizocarp. Fruit capsule or berry or nut

→ Seeds - endospermic

Genera included under Aristolochieae
- → Apama
- → Aristolochia
- → Asarum
- → Euglypha
- → Holostylis
- → Saruma
- → Thottea

Sub class: Monochlamydeae

Series: microembryae

Families:

Piperaceae

Chloranthaceae

Myristiceae

Moniaceae

Piperaceae

Classification (Bentham and Hooker)

- Phanerogams
- Dicotyledons
- Monochlamydeae
- Achlamydosporae
- Piperaceae

General characters

→ Shrubs, herbs and very rarely trees., often climbing.
→ Stem - generally succulent and herbaceous. They are often provided with jointed or swollen nodes. Stem possesses distinct and scattered vascular bundles like that of monocotyledons. Bundles are open. Resin and oil secreting sacs are generally found both in the epidermis and ground tissue.
→ Leaves - alternate, rarely opposite or whorled, petiolate, entire, stipulate or exstipulate; when stipules are present, they are adnate to the
→ Inflorescence - Catkin like terminal or axillary spikes.
→ Sometimes spikes are found to be situated opposite to the leaves (eg:- Piper)
→ Flowers - naked which are often more or less sunk in the fleshy axis of the spike. Flowers are

Minute bracteate, usually hermaphrodite rarely unisexual.
→ Perianth - altogether absent. Flowers naked
→ Androecium - Stamen number varies from 1 to 10. Mostly trimerous with 2 whorls, each with 3 stamens. In *Piper nigrum* there are only 2 stamens, the posterior stamen of the inner whorl has been aborted. Sometimes the whole inner whorl is found to be absent. This arrangement of stamen is a constant feature and has been considered to be a constantly reduced form. Filaments distinct, anthers bicelled and dehisce by longitudinal slits.
→ Gynoecium - carpel number varies from 2-5. Ovary- superior, one chambered with a solitary basal orthotropous ovule. Style absent or one. Stigmas are as many as the number of carpels. In *Peperomia*, stigma is simple often brushlike and lateral.
→ Fruit is a small drupe
→ Seeds - small, endospermic with minute embryo.

Genera included under Peperaceae

→ Circaeocarpus
→ Lindeniopiper
→ Ottonia
→ Piper
→ Pothomorphe
→ Sarcorhachis
→ Trianaeopiper
→ Zippelia

Chapter - 140
Chloranthaceae

Classification (Bentham and Hooker)

Phanerogams
Dicotyledons
Monochlamydeae
Micrombryae
Chloranthaceae

General characters

→ Trees, shrubs and herbs.; Annual or Perennial; leaves - opposite, petiolate, connate, aromatic, simple, epulvinate, stipulate. Stipules interpetiolar; lamina margin - serrate

→ Plants hermaphrodite or dioecious.

→ Inflorescence - spikes/heads/panicles.

→ Perianth sepaline - (female and hermaphrodite flowers) or absent (male flowers); when present - 3; joined, 1 whorled

→ Androecium - 1 to 5 stamens, free of the perianth, united with gynoecium; coherent; monadelphous; laminar or filantherous; Anthers dehiscing via longitudinal slits or valves; unilocular to bilocular;

→ Gynoecium - 1 carpelled, monomerous, superior or partly inferior; Placentation - apical; Ovules pendulous
→ Fruit - fleshy, indehiscent, drupaceous.
→ Seeds - Endospermic

Genera included under chloranthaceae
→ Ascarina   → Chloranthus   → Hedyosmum
→ Sarcandra

Chapter - 141
Myristiceae

Classification (Bentham and Hooker)

Phanerogams
Dicotyledons
Monochlamydeae
Microembryae
Myristiceae

General characters

→ Trees, with coloured juice (red sap); bearing essential oils.
→ Leaves - Persistent, alternate, spiral to distichous; leathery, Petiolate, non-sheathing; often gland dotted; often aromatic, simple; lamina - entire, pinnately veined; exstipulate
→ Plants monoecious or dioecious
→ Female flowers without staminodes; gynoecium of male flowers absent
→ Inflorescence - cymes/ fascicles/ racemes/ heads
→ Flowers - bracteate, trimerous, cyclic
→ Perianth sepaline; joined, 1 whorled. Calyx 2-5 sepals, 1 whorled, gamosepalous, blunt-lobed valvate.

→ Androecium 2-30 stamens; free of perianth; coherent (monadelphous); 1 whorled; Anthers separate or coherent; extrose; dehiscing via longitudinal slits.

→ Gynoecium - 1 carpelled; monomerous; superior; non-stylate or stylate (subsessile); apically stigmatic; 1 ovuled; Placentation basal; anatropous.

→ Fruit - fleshy to non-fleshy; dehiscent - legume

→ Seeds - endospermic

Genera included under Myristicaea

- Bicuiba
- Brochoneura
- Cephalosphaera
- Coelocaryon
- Compsoneura
- Doyleanthus
- Endocomia
- Gymnacranthera
- Haematodendron
- Horsfieldia
- Iryanthera
- Knema
- Mauloutchia
- Myristica
- Osteophloeum
- Otoba
- Pycnanthus
- Scyphocephalium
- Staudtia
- Virola.

Sub class: Monochlamydeae

Series: Daphnales

Families:

Laurineae

Proteaceae

Thymelaceae

Penaeaceae

Eleagnaceae

Chapter - 143
Laurineae

Classification (Bentham and Hooker)

Phanerogams
Dicotyledons
Monochlamydeae
Daphnales
Laurineae

General characters

→ Trees and shrubs
→ Leaves - Persistent, usually alternate rarely opposite or whorled; spiral, leathery, petiolate; non-sheathing; gland dotted, aromatic, simple. Lamina - entire; exstipulate.
→ Plants hermaphrodite or monoecious or dioecious.
→ Inflorescence - cymose or racemose. Rarely solitary.
→ Flowers - often fragrant, regular, trimerous, cyclic; free hypanthium present. Hypogynous disk present or absent.
→ Perianth with distinct calyx and corolla or sepaline or of tepals; 6 or 4, free; 1-3 whorled isomerous, sepaloid to petaloid, green/white/cream/yellow. Fleshy or non-fleshy.

Calyx if present 4-6 sepals in 2 whorl, Polysepalous, regular, imbricate
→ Corolla consists of 3 petals, 1 whorled, polypetalous, imbricate, sessile.
→ Androecium 2-26 stamens; free of the perianth, 1-3 whorled; diplostemonous to polystemonous; filantherous. Anthers basifixed, introrse, dehiscing by longitudinal valves.
→ Gynoecium - Mono or tricarpellary. Superior or inferior; Style - apical; Placentation apical
→ Fruit - fleshy or rarely non-fleshy; indehiscent, drupaceous or baccate. Fruit enclosed in a fleshy receptacle or enclosed in the fleshy hypanthium or without fleshy investment.
→ Seeds non-endospermic.

Genera included under Laurineae

- Actinodaphne
- Adenodaphne
- Aiouea
- Alseodaphne
- Anaueria
- Aniba
- Apollonias
- Aspidostemon
- Beilschmiedia
- Brassiodendron
- Caryodaphnopsis
- Chlorocardium
- Cinnadenia
- Cinnamomum
- Clinostemon
- Cryptocarya
- Dehaasia
- Dicypellium
- Dodecadenia
- Endiandra
- Endlicheria
- Eusideroxylon
- Gamanthera
- Hexapora
- Hypodaphnis
- Iteadaphne
- Kubitzkia
- Laurus
- Licaria
- Lindera
- Litsea
- Machilus
- Mezilaurus
- Nectandra
- Neocinnamomum
- Neolitsea
- Nothaphoebe
- Ocotea
- Paraia
- Persea
- Phoebe
- Phyllostemonodaphne
- Pleurothyrium
- Potameia
- Potoxylon
- Povedadaphne
- Ravensara
- Rhodostemonodaphne
- Sassafras
- Syndiclis
- Triadodaphne
- Umbellularia
- Urbanodendron
- Williamodendron

Chapter – 144
Proteaceae

Classification (Bentham and Hooker)

Phanerogams
Dicotyledons
Monochlamydeae
Daphnales
Proteaceae

General Characters

→ Trees or shrubs or herbs
→ Leaves – alternate, opposite or whorled; usually spiral, flat or terete; leathery or fleshy or modified into spines; petiolate to sessile; occasionally aromatic; simple, dissected or entire; often accular or linear; exstipulate. Margin entire/crenate/serrate/dentate.
→ Plants hermaphrodite or monoecious or dioecious
→ Flowers in inflorescence – racemes/spikes/heads/umbers; solitary, axillary.
→ Flowers – bracteate or ebracteate, cyclic; floral receptacle developing a gynophore; free hypanthium present (represented by a

calyx tube with stamens attached) or absent
→ Perianth with distinct calyx and corolla (glands or scales)
→ Calyx consists of 4 sepals, gamosepalous, 3 joined & 1 free. Calyx tubular, valvate
→ Corolla consists of 2-4 petals (represented by glands or scales)
→ Androecium - 4 stamens, free of the perianth or adnate, 1 whorled; anthers basifixed, strongly introrse, dehiscing via longitudinal slits.
→ Gynoecium - monocarpellary, superior, Placentation mostly marginal or apical; ovary sessile to stipitate. Styles bearing an 'indusium' beneath the stigma.
→ Fruit - fleshy or non fleshy; dehiscent or indehiscent; follicle or drupaceous or nucular or achene
→ Seeds - non endospermic or endospermic

Genera included under Proteaceae

- Acidonia
- Adenanthos
- Agastachys
- Alloxylon
- Athertonia
- Aulax
- Austromuellera
- Banksia
- Beauprea
- Beaupreopsis
- Bellendina
- Brabejum
- Buckinghamia
- Cardwellia
- Carnarvonia
- Cenarrhenes
- Conospermum
- Darlingia
- Diastella
- Dilobeia
- Dryandra
- Embothrium
- Eucarpha
- Euplassa
- Faurea
- Finschia
- Floydia
- Franklandia
- Garnieria
- Gevuina
- Grevillea
- Hakea
- Helicia
- Helicopsis
- Hicksbeachia
- Hollandaea
- Isopogon
- Kermadecia
- Knightia
- Lambertia
- Leucadendron
- Leucospermum
- Lomatia
- Macadamia
- Malagasia
- Mimetes
- Musgravea
- Neorites
- Opisthiolepis
- Oreocallis
- Orites
- Orothamnus
- Panopsis
- Paranomus
- Persoonia
- Petrophile
- Placospermum
- Protea
- Roupala
- Serruria
- Sleumerodendron
- Sorocephalus
- Spatalla
- Sphalmium
- Stenocarpus
- Stirlingia
- Strangea
- Symphionema
- Synaphea
- Telopea
- Toronia
- Triunia
- Turrillia
- Vexatorella
- Virotia
- Xylomelum

Chapter - 145
Thymelaceae

Classification (Bentham and Hooker)

Phanerogams
Dicotyledons
Monochlamydeae
Daphnales
Thymelaceae

General Characters

- Bark usually shiny and fibrous
- Number of stamens is usually once or twice the number of calyx lobes
- If twice then in 2 whorls
- The floral tube appears to be calyx or corolla, but is actually a hollow receptacle
- The sepals are mounted on the rim of the floral tube
- Stamens may be mounted on the rim or inside
- The petals are actually stipular appendages of the sepals
- Fruit - 1 seeded berry or achene

Families of Dicotyledons

Genera included under Thymelaeaceae

- Aetoxylon
- Amyxa
- Aquilaria
- Arnhemia
- Atemnosiphon
- Craterosiphon
- Dais
- Daphne
- Daphnopsis
- Deltaria
- Dicranthron
- Dicranolepis
- Dirca
- Drapetes
- Edgeworthia
- Englerodaphne
- Enkleria
- Eriosolena
- Funifera
- Gnidia
- Gonystylus
- Goodallia
- Gyrinops
- Jedda
- Kelleria
- Lachnaea
- Lagetta
- Lasiadenia
- Lethedon
- Linodendron
- Linostoma
- Lophostoma
- Octolepis
- Ovidia
- Passerina
- Peddiea
- Phaleria
- Pimelea
- Rhamnoneuron
- Schoenobiblus
- Solmsia
- Stellera
- Stephanodaphne
- Struthiola
- Synandrodaphne
- Synaptolepis
- Tepuianthus
- Thecanthes
- Thymelaea
- Wikstroemia

Chapter - 146
Penaeaceae

Classification (Bentham and Hooker)

Phanerogams
Dicotyledons
Monochlamydeae
Daphnales
Penaeaceae

General characters

→ Shrubs
→ Leaves - decussate, simple, entire, glabrous, linear to orbicular.
→ Inflorescence highly variable
→ Flowers - sessile or pedicellate, bisexual, actinomorphic, tetramerous, perigynous; sepals - free, petaloid, triangular to ovate, persistent.
→ Stamens as many as alternating the sepals; free, inserted on the rim of the hypanthium; basifixed, introrse with longitudinal dehiscence.
→ Pistil - 4 carpellate, syncarpous, superior, 4-locular; stigma terminal, capitate, style winged; placentation - axile;

anatropous
→ Fruit - loculicidal capsule
→ Seeds - ovoid, slightly compressed; brown to almost black when mature.

Genera included under penaeaceae

- → Brachysiphon
- → Endonema
- → Glischrocolla
- → Penaea
- → Saltera
- → Sonderothamnus
- → Stylapterus

Chapter - 147
Eleagnaceae

Classification (Bentham and Hooker)

- Phanerogams
- Dicotyledons
- Monochlamydeae
- Daphnales
- Eleagnaceae

General characters

→ Commonly thorny with simple leaves often coated with tiny scales or hairs. Most are xerophytes, some are halophytes

→ Elaeagnaceae members often harbour nitrogen fixing actinomycetes of the genus Frankia in root nodules.

→ Stems and leaves are covered with silvery brown or golden hairs which are either peltate or scaly.

→ Unisexual, dioecious

→ Petals absent; Perianth consisting of single whorl of 2 to 8 fused sepals

→ In male flowers, receptacle is often flat, while in bisexual and female flowers it is tubular.

→ 4 to 8 stamens with free filaments and

bilocular anthers.
- Ovary superior with one carpel containing a single, erect anatropous ovule
- Style long and bears a single stigma
- Fruit - achene or drupe, enclosed by the thickened lower part of the persistent calyx.
- Single seeded, non-endospermic with a straight embryo and fleshy cotyledon.

Genera included under Eleagnaceae
- Elaeagnus - Hippophae - Shepherdia

Sub class: Monochlamydeae

Series: Achlamydosporeae

Families:

Loranthaceae

Santalaceae

Balanophoreae

Chapter-148
Loranthaceae

Classification (Bentham and Hooker)
- Phanerogams
- Dicotyledons
- Monochlamydeae
- Achlamydosporeae
- Loranthaceae

General Characters

→ Either herbs or shrubs. They live on tree branches in semiparasitic state. They remain attached to their hosts by suckers.

→ Roots represented by suckers or haustoria which derive their nourishment from host tree. Such roots branch within host tissue.

→ Stem - nodes swollen; dichotomously branched, woody, glabrous, solid, cylindrical, green in colour.

→ Leaves - Simple, entire, persistent, coriaceous, thick, opposite or whorled, rarely alternate, exstipulate, glabrous. Sometimes leaves are very much reduced and represented by scales.

→ Inflorescence - Solitary or may be arranged in dichasia. Each flower remains situated in the axil of a bract. Also - raceme, spike or panicle

→ Flowers are actinomorphic with a tendency to zygomorphy. Either hermaphrodite / unisexual and dioecious., incomplete and epigynous, small and green.; dimerous or trimerous in their plan.

→ Perianth – there is a cupular receptacle and from its edge a perianth of generally two similar 2-3 merous whorls arises. Perianth leaves may be free or united. In subfamily Loranthoideae, they are large and brilliantly coloured (petaloid). A slightly toothed or irregular rim known as calyculus is present below the perianth and this may be treated as calyx.

→ Androecium – Number of stamens is equal to the segments of the perianth and situated opposite the perianth leaves. They are more or less united with the perianth. In Viscum, this union is incomplete, here the petals and stamens arise as one structure from the floral axis on the front of which a large number of pollen sacs are produced.

→ Gynoecium consists of 3-4 syncarpous carpels. Ovary-inferior, unilocular with a large central placenta having many ovules. Ovules are not differentiated from the placenta. Placentation -basal

→ Fruit is a pseudocarp, the ovary unites with the receptacular cup developing a berry like or more rarely a drupaceous fruit. Fruit may be compared to that of apple.

→ Seeds - endospermic, embryo is straight.

→ Pollination is generally anemophilous, but in Loranthoideae due to brightly coloured flowers pollination is entomophilous.

Genera Included under Loranthaceae

- Actinanthella
- Aetanthus
- Agelanthus
- Alepis
- Amyema
- Amylotheca
- Atkinsonia
- Bakerella
- Baratranthus
- Benthamina
- Berhautia
- Cecarria
- Cladocolea
- Cyne
- Dactyliophora
- Decaisnina
- Dendropemon
- Dendrophthoe
- Desmaria
- Diplatia
- Distrianthes
- Elytranthe
- Emelianthe
- Englerina
- Erianthemum
- Gaiadendron
- Globimetula
- Helicanthes
- Helixanthera
- Ileostylus
- Ixocactus
- Kingella
- Lampas
- Lepeostegeres
- Lepidaria
- Ligaria
- Loranthus
- Loxanthera
- Lysiana
- Macrosolen
- Moquinella
- Muellerina
- Notanthera
- Nuytsia
- Oliverella
- Oncella
- Oncocalyx
- Oryctanthus
- Tupeia
- Oryctina
- Panamanthus
- Papuanthes
- Pedistyles
- Peraxella
- Phragmanthera
- Phthirusa
- Plicosepalus
- Psittacanthus
- Scurrula
- Septulina
- Socratina
- Sogerianthe
- Spragueanella
- Struthanthus
- Tapinanthus
- Taxillus
- Tetradyas
- Thaumasianthes
- Tolypanthus
- Trileprdea
- Tripodanthus
- Tristerix
- Trithecanthera
- Vanwykia

Chapter - 149
Santalaceae

Classification (Bentham and Hooker)

Phanerogams
Dicotyledons
Monochlamydeae
Achlamydosporeae
Santalaceae

General Characters

→ Semi Parasitic herbs, shrubs or small trees.
→ They remain attached to the hosts by means of haustoria. The haustoria penetrate the host's root as far as xylem, as a result the vessel of the host and that of parasite are organically jointed, the inter communicating vascular elements are known as 'phloeotracheids'. The haustoria may be simple or compound. The structure of a mature haustorium is like an inverted flask.
→ Stem - erect, branched, solid, cylindrical, glabrous green when young, brown or grey when old
→ Leaves - simple, entire, leathery, opposite, sometimes alternate, exstipulate; acute apex, petiolate.
→ Inflorescence — Solitary or arranged in dichasia
→ Flower — Small, actinomorphic, haplochlamydous, bisexual or unisexual by abortion; Pentamerous or

trimerous; Nectar secreting disc present in many genera epigynous or hypogynous.

→ Perianth – Tepals 4-5 or 3-6 united at the base; Valvate aestivation; either sepaloid or petaloid; Pentamerous (Comandra), others – trimerous.

→ Androecium – Stamens are as many as perianth members and opposite to them and are epiphyllous (adnate to perianth leaves); filaments – short, anthers dithecous, dehisce vertically.

→ Gynoecium – carpels 3-5 rarely 2, united in unilocular inferior ovary or semi-inferior; ovules 3; orthotropous to anatropous; only one fertile borne on the apical end of central placenta and pendulous; seeds without testa and with a fleshy endosperm. Ovule is without integument, hence seed without testa. _Exocarpus_ has semi inferior naked ovary having single sessile naked ovule with pollen chamber resembling Gymnosperm.

→ Fruit – an achene or drupe.

Genera included under Santalaceae

- Acanthosyris
- Amphorogyne
- Antholobus
- Arjona
- Austroamericium
- Buckleya
- Cervantesia
- Choretrum
- Cladomyza
- Colpoon
- Comandra
- Daenikera
- Dendromyza
- Dendrotrophe
- Dufrenoya
- Elaphanthera
- Exocarpos
- Geocaulon
- Jodina
- Kunkeliella
- Leptomeria
- Mida
- Myoschilos
- Nanodea
- Nestronia
- Okoubaka
- Omphacomeria
- Osyridocarpus
- Osyris
- Phacellaria
- Pyrularia
- Quinchamalium
- Rhoiacarpus
- Santalum
- Scleropyrum
- Spirogardnera
- Thesidium
- Thesium

Sub class: Monochlamydeae

Series: Unisexuales

Families:

Euphorbiaceae

Balanopseae

Urticaceae

Plantanaceae

Leitnerieae

Juglandeae

Myricaceae

Casuarineae

Cupuliferae

Chapter – 157
Euphorbiaceae

Classification (Bentham and Hooker)

Phanerogams
Dicotyledons
Monochlamydeae
Unisexuales
Euphorbiaceae

General characters:

→ Plants are herbs, shrubs or trees. Species of *Tragia* are climbers.

→ Majority of the members of the family possess large lateciferous vessels which contain latex.

→ Branched tap root system.

→ Stem – herbaceous or woody, erect, very rarely climbing. *Xylophylla* possess flat phylloclades. Stem is branched. May be cylinderical, angular or flat. Usually solid but in *Ricinus communis* it is hollow. Many stem possess spines.

→ Leaves – usually alternate but sometimes opposite (*Euphorbia hirta*). Usually leaves are simple but rarely deeply incised (*Ricinus*, *Manihot*). Leaves of many *Euphorbia* sp are scaly and caducous. In many cases leaves are reduced to spines.

In some species leaves are reduced by cladodes. Usually the leaves are stipulate; stipules become branched and hair like; stipules represented by glands or spines (Euphorbia)

→ Inflorescence varies greatly. May be racemose or cymose or complex. In Euphorbia, inflorescence is peculiar and known as cyathium (cyme modification). Cyathium inflorescence – a large number of male flowers each represented by a stalked stamen are found arranged around a central stalked female flower. Female flower consists of Gynoecium only. The complete inflorescence looks like a single flower. The bracts are being arranged like a perianth. Bracts united to form a cup like structure.

Acalypha – inflorescence is catkin type.
Croton & Ricinus – flowers in terminal racemes.
Jatropha – cymose type and flowers arranged in terminal cymes.
Manihot – Racemes.

→ Flowers are always unisexuals. Much reduced; may be monoecious or dioecious; incomplete, regular, actinomorphic and hypogynous.

→ Perianth is distinguished into calyx and corolla in some cases (Croton). In majority of the species Perianth is absent. <u>Ricinus</u> – calyx is present, corolla absent. <u>Euphorbia</u> – Perianth is absent or represented by tiny scaly structures.; Perianth has 4 to 5 petals, calyx and corolla consists of 4 to 5 sepals or petals. Aestivation – valvate or imbricate.

→ Androecium – number of stamens 1–∞. Usually the stamens number equals the perianth leaves. In Euphorbia, a single stalked stamen represents a single male flower. <u>Ricinus</u>, 5 profusely branched stamens are present. <u>Jatropha</u> stamens are arranged in 2 whorls of 5. In many stamens are indefinite (<u>Croton</u>). Filaments free or united. Anthers are dithecous and dehisce by apical pores or transverse or longitudinal slits.

→ Gynoecium – 3 carpels (tricarpellary), syncarpous, ovary – trilocular, superior; each locule has 1 or 2 pendulous, anatropous ovules. Placentation axile.

→ Fruit – Schizocarpic. Fruit break violently and dehisce into one seeded cocci. Such fruit is called regma.

→ Seeds – endospermic.

→ Pollination entomophilous.

Genera Included under Euphorbiaceae

- Acalypha
- Acidocroton
- Acidoton
- Actephila
- Adelia
- Adenochlaena
- Adenocline
- Adenopeltis
- Adenophaedra
- Adriana
- Aerisilvaea
- Afrotrewia
- Agrostistachys
- Alchornea
- Alchorneopsis
- Aleurites
- Algernonia
- Alphandia
- Amanoa
- Amperea
- Amyrea
- Andrachne
- Angostyles
- Annesijoa
- Anomalocalyx
- Anthostema
- Aparisthium
- Apodiscus
- Aporusa
- Argomuellera
- Argythamnia
- Aristogeitonia
- Ashtonia
- Astrocasia
- Astrococcus
- Austrobuxus
- Avellanita
- Baccaurea
- Baliospermum
- Baloghia
- Benoistia
- Bernardia
- Beatya
- Beyeria
- Blachia
- Blotia
- Blumeodendron
- Bocquillonia
- Bonania
- Borneodendron
- Bossera
- Botryophora
- Breynia
- Bridelia
- Calycopeplus
- Canaca
- Caperonia
- Caryodendron
- Caslitoa
- Calvacoa
- Celanodendron
- Celpanella
- Cephalocroton
- Cephalomappa
- Chaetocarpus
- Chascotheca
- Cheilosa
- Chiropetalum
- Chlamydojatropha
- Chondrostylis
- Chonocentrum
- Choriceras
- Chrozophora
- Cladogelonium
- Cladogynos
- Claoxylon
- Claoxylopsis
- Cleidiocarpon
- Cleidion

- Cleistanthus
- Clutia
- Cnesmone
- Cnidoscolus
- Cocconerion
- Codiaeum
- Colliguaja
- Conceveiba
- Cordemoya
- Croizatia
- Croton
- Crotonogyne
- Crotonogynopsis
- Crotonopsis
- Ctenomeria
- Cubanthus
- Cyrtogonone
- Cyttaranthus
- Dalechampia
- Dalembertia
- Deuteromallotus
- Deutzianthus
- Dichostemma
- Dicoelia
- Didymocistus
- Dimorphocalyx
- Discocarpus
- Discoclaoxylon
- Dicoelodeon
- Discoglypremna
- Dissiliaria
- Ditaxis
- Ditta
- Dodecastigma
- Domohinea
- Doryxylon
- Droceloncia
- Drypetes
- Duvigneaudia
- Dysopsis
- Elaeophorbia
- Elateriospermum
- Eleutherostigma
- Endadenium
- Endospermum
- Enriquebeltrania
- Epiprinus
- Eremocarpus
- Erismanthus
- Erythrococca
- Euphorbia
- Excoecaria
- Fahrenheitia
- Flueggea
- Fontainea
- Garcia
- Gavarretia
- Givotia
- Glochidion
- Glycydendron
- Glyphostylus
- Grimmeodendron
- Grossera
- Gymnanthes
- Haematostemon
- Hamilcoa
- Hevea
- Heywoodia
- Hippomane
- Homonoia
- Hura
- Hyaenanche
- Hieronima
- Hylandia
- Jablonskia
- Jatropha
- Joannesia
- Kairothamnus

- Keayodendron
- Klaineanthus
- Koilodepas
- Lachnostylis
- Lasiococca
- Lasiocroton
- Lautembergia
- Leeuwenbergia
- Leidesia
- Leptonema
- Leptopus
- Leucocroton
- Longetsheimia
- Lobanilia
- Loerzingia
- Longetia
- Mabea
- Macaranga
- Maesobotrya
- Mallotus
- Manihot
- Manihotoides
- Mannilophyton
- Maprounea
- Mareya
- Moacroton
- Monodenium
- Monotaxis
- Moultonianthus
- Myladenia
- Myricanthe
- Nealchornea
- Necepsia
- Neoboutonia
- Neoguillauminia
- Neoholstia
- Neoroepera
- Neoscortechinia
- Neotrewia
- Octospermum
- Oldfieldia
- Oligoceros
- Omalanthus
- Omphalea
- Omphellantha
- Opthalmoblapton
- Oreoporanthera
- Ostodes
- Pachystroma
- Pachystyledium
- Pantadenia
- Paradrypetes
- Paranecepsia
- Parapantadenia
- Parodiodendron
- Paussandra
- Pedilanthus
- Pentabrachion
- Petalodiscus
- Petalostigma
- Philyra
- Phyllanoa
- Phyllanthus
- Piryelodendron
- Piranhea
- Plageostyles
- Platygyna
- Plukenetia
- Podadenia
- Podocalyx
- Pogonophora
- Poilaniella
- Poinsettia
- Polyandra
- Poranthera
- Protomegabaria
- Pseudagrostistachys
- Pseudanthus
- Pseudocroton

- Pseudolachnostylis
- Pterococcus
- Ptychopyxis
- Putranjiva
- Pycnocoma
- Reutealis
- Reverchonia
- Richeria
- Richeriella
- Ricinocarpus
- Ricinodendron
- Ricinus
- Rockinghamia
- Romanoa
- Sagotia
- Sampantea
- Sandwithia
- Sapium
- Sauropus
- Savia
- Scagea
- Schizzophyton
- Sebastiania
- Securinega
- Seidelia
- Senefeldera
- Senefelderopsis
- Sibangea
- Spathiostemon
- Sperankia
- Sphaerostylis
- Sphysanthera
- Spirostachys
- Spondianthus
- Stachyandra
- Stachystemon
- Strillingia
- Strophioblachia
- Sumbaviopsis
- Suregada
- Symphyllia
- Synandenium
- Syndyophyllum
- Tacarung
- Tannodia
- Tapoides
- Tetracoccus
- Tetraplandra
- Tetrorchidium
- Thecacoris
- Thyrsanthera
- Tragia
- Tragiella
- Trevia
- Trigonopleura
- Trigonostemon
- Vaupesia
- Veanisia
- Vigea
- Voatamalo
- Wetria
- Whyanbeelia
- Wielandia
- Zimmermannia
- Zimmermanniopsis

Chapter - 150
Balanophoreae

Classification (Bentham and Hooker)

Phanerogams
Dicotyledons
Monochlamydeae
Achlamydosporeae
Balanophoreae

General Characters

→ Small to tall evergreen trees
→ Leaves - alternate, dimorphic, shoots with minute scale leaves; petiolate; leaves petiolate;
→ Inflorescence - catkin
→ vestigial perianth
→ Stamens 1-12, filaments short; anthers dehisce laterally
→ Numerous densely crowded, spirally arranged deltoid 'cupular' bracts subtending the naked ovary; syncarpous; 2-3 carpellary, ovary 2-3 locular; ovules 2 per carpel; axile placentation
→ Fruit - drupe
→ Seeds endospermic

Genera included under Balanophoreae

- Balanophora
- Chlamydophytum
- Corynaea
- Dactylanthus
- Ditepalanthus
- Exorhopala
- Hachettea
- Helosis
- Langsdorffia
- Lophophytum
- Mystropetalon
- Ombrophytum
- Rhopalocnemis
- Sarcophyte
- Scybalium
- Thonningia

Chapter-15.3
Urticaceae

Classification (Bentham and Hooker)

- Phanerogams
- Dicotyledons
- Monochlamydeae
- Unisexuales
- Urticaceae

General characters

→ Shrubs, lianas, herbs or few trees

→ Leaves - small to large; alternate or opposite; sessile (rarely); usually petiolate; sessile, non-sheathing; simple, lamina - entire (usually); stipulate or exstipulate; Lamina margin - entire/ serrate/ dentate.

→ Plants monoecious or dioecious. Female flowers with or without staminodes; Gynoecium of male flowers vestigial or absent

→ Flowers rarely solitary; inflorescence — spike, raceme, panicle, cyme; with or without involucral bracts.

→ Flowers - bracteate or ebracteate, minute or small; regular; 3-6 merous;

→ Perianth - sepaline or vestigial to absent; calyx 2-6 (in males) or 3-5 (in females); 1 whorled; polysepalous or gamosepalous;

regular, persistent; imbricate or valvate
→ Androecium 2-6 stamens; free; 1 whorled; stamens isomerous with perianth; Anthers dorsifixed; dehiscing via longitudinal slits; introrse
→ Gynoecium - monocarpellary, monomerous or syncarpous; superior or partly inferior; style short or absent; 1 ovule; placentation basal; stigma -1; ovules in single cavity.
→ Fruit - fleshy or non fleshy; indehiscent; an achene, or nucular or drupaceous; syncarpous, indehiscent; achene like or nut or drupe
→ Seeds - scantily endospermic or non-endospermic

Genera included under Urticaceae

- Aboriella
- Achudemia
- Archiboehmeria
- Astrothalamus
- Australina
- Boehmeria
- Chamabainia
- Cypholophus
- Debregeasia
- Dendrocnide
- Didymodoxa
- Discocnide
- Droguetia
- Elatostema
- Forsskaolea
- Gesnouinia
- Gibbsia
- Girardinia
- Laportea
- Lecanthus
- Leucosyke
- Maoutia
- Meniscogyne
- Myriocarpa
- Nanocnide
- Neodistemon
- Neraudia
- Nothocnide
- Obetia
- Oreocnide
- Parietaria
- Pellionia
- Petelotiella
- Phenax
- Pilea
- Pipturus
- Pouzolzia
- Procris
- Rousselia
- Sarcochlamys
- Sarcopilea
- Soleirolia
- Touchardia
- Urera
- Urtica

Chapter - 154
Plantanaceae

Classification (Bentham and Hooker)
- Phanerogams
- Dicotyledons
- Monochlamydeae
- Unisexuales
- Plantanaceae

General Characters

→ Large trees

→ Leaves - deciduous, medium sized; large or alternate, spiral, flat, petiolate, sheathing, simple. Stipulate; stipules - ochreate, scaly, caducous; Margin - dentate

→ Plants monoecious; female flowers with staminodes, gynoecium of male flowers - vestigial or absent.

→ Pollination - anemophilous.

→ Inflorescence - heads

→ Flowers - bracteate or ebracteate; small; regular; cyclic.

→ Perianth with distinct calyx and corolla or sepaline. Calyx consists of 3-7 sepals; 1 whorled; poly or gamosepalous; regular. Corolla in male flowers → 3 to 7 petals; 1 whorled;

Polypetalous.

→ Androecium - 3 to 7 stamens; 1 whorled; isomerous with perianth; filantherous; Anthers - basifixed or adnate; dehiscing by longitudinal slits; latrose.

→ Gynoecium consists of 3-9 carpels; isomerous with perianth; apocarpous; superior; apically stigmatic; 1-2 ovuled; placentation apical to marginal; ovules pendulous;

→ Fruit- non fleshy, aggregate; indehiscent - achene/nucular. Gynoecia of adjoining flowers combine to form a multiple fruit

→ Seeds scantily endospermic.

Genera included under Plantanaceae
- Ambiplatanus
- Ettingshuausenia
- Macginicarpa
- Macginitiea
- Plataninium
- Platanus

Chapter-155
Leitnerieae

Classification (Bentham and Hooker)

- Phanerogams
- Dicotyledons
- Monochlamydeae
- Unisexuales
- Leitnerieae

General Characters

→ Small trees or shrubs; resinous;
→ Leaves – deciduous, alternate, petiolate, simple lamina – entire; elliptic-oblong or lanceolate; cuneate at the base; exstipulate; Margin – entire
→ Plants mostly dioecious; Gynoecium of male flowers absent; Pollination – anemophilous
→ Inflorescence – Catkins, cymose
→ Flowers – bracteate; mostly bracteolate; small.
→ Perianth sepaline (female flowers) or absent (male flowers).
→ Calyx (Perianth, here alternatively as bracts or bracteoles); 3-8 sepals; gamosepalous.
→ Androecium consists of 1-5 or 3-15 stamens; free of one another; stamens shortly

filantherous; Anthers basifixed; dehiscing via longitudinal slits.
→ Gynoecium – mono or bicarpellary; syncarpous; superior. Style-1, apical; Stigma-1. Placentation Parietal; 1 ovule per cavity.
→ Fruit – fleshy to non-fleshy; indehiscent; drupaceous.
→ Seeds thinly endospermic

Genera included under Leitnerieae
→ Leitneria

Chapter - 156
Juglandeae

Classification (Bentham and Hooker)

Phanerogams
Dicotyledons
Monochlamydeae
Unisexuales
Juglandeae

General characters

→ Mostly trees or few shrubs;
→ Leaves - deciduous, mostly alternate or opposite, petiolate or sessile; non-sheathing; aromatic; compound; exstipulate
→ Plants monoecious / dioecious; Gynoecium of male flowers vestigial or absent; pollination anemophilous.
→ Inflorescence - catkins / racemes / spikes. rarely solitary;
→ Flowers - bracteate; 2 bracteolate or absent
→ Perianth - Sepaline or vestigial or absent; calyx consists of 1-5 sepals; more or less adnate with the bracteoles or obsolete in male flowers, consisting of 4 calyx teeth or suppressed in female flowers.

→ Androecium - 3 to many stamens; free of the Perianth; diplo to Polystemonous; Shortly filantherous; Anthers - basifixed; dehiscing via longitudinal slits.

→ Gynoecium 2-3 carpelled, Syncarpous, inferior ovary - uni or bi or tri locular. Locules may be sub divided by false septa; styles - 1 + 2, apical; stigma - 2; placentation - basal

→ Fruit - fleshy or non fleshy; indehiscent; drupe, or nut or Samara.

→ Seeds - non endospermic

Genera included under Juglandaceae

→ Alfaroa
→ Carya
→ Cyclocarya
→ Engelhardtia
→ Juglans
→ Oreomunnia
→ Platycarya
→ Pterocarya

Chapter – 157
Myricaceae

Classification (Bentham and Hooker)

Phanerogams
Dicotyledons
Monochlamydeae
Unisexuales
Myricaceae

General characters

→ Trees or shrubs; resinous
→ Leaves – persistent or deciduous; alternate; spiral; petiolate, non-sheathing; aromatic, simple.
→ Lamina – entire or dissected, usually exstipulate
   Margin – entire or serrate.
→ Plants monoecious or dioecious.
→ Inflorescence – spikes
→ Flowers – bracteate, bracteolate; cyclic; In male flowers hypogynous disk is present whereas in female flowers it is absent.
→ Perianth absent
→ Androecium – 2–6 stamens, free of one another or filaments sometimes connate; Monadelphous or 1 whorled; Anthers – extrorse, dehiscing via longitudinal slits.

→ Gynoecium - bicarpellary, syncarpous, superior or partly inferior; ovary unilocular; styles - 2; free to partially joined, apical; stigma - 2. Placentation basal.
→ Fruit - fleshy to non fleshy, indehiscent, drupe
→ Seeds - scantily endospermic or non-endospermic

Genera included under Myricaceae
→ Canacomyrica → Comptonia → Gale → Myrica.

Chapter - 159
Cupuliferae

Classification (Bentham and Hooker)

Phanerogams
Dicotyledons
Monochlamydeae
Unisexuales
Cupuliferae

General characters

→ Trees and shrubs
→ Leaves - alternate, spiral or distichous, herbaceous or leathery; non sheathing; simple; epulvinate; lamina dissected or entire; stipulate; Margin entire.
→ Plants monoecious, rarely dioecious
→ Pollination anemophilous or entomophilous
→ Inflorescence - Catkins or flowers solitary, axillary
→ Flowers - bracteate, minute.
→ Perianth - sepaline or vestigial. Calyx 4-7 sepals Poly or gamosepalous; blunt lobed, regular, imbricate
→ Androecium consists of 4-40 stamens; free of one another, one whorled; Anthers - dehisc

via longitudinal slits. Stamens – isomerous with the perianth to diplostemonous.

→ Gynoecium 2-3 carpelled or rarely 6-12 carpelled; syncarpous, inferior. Styles usually 2-3 rarely 6-12; placentation axile or apical. Ovules 2 per locule; anatropous

→ Fruit – non fleshy, indehiscent, nut or samara
→ Seeds – non-endospermic

Genera included under Cupuliferae

- Fagus
- Nothofagus
- Lithocarpus
- Castanopsis
- Colombobalanus
- Castanea
- Chrysolepis
- Quercus
- Trigonobalanus

Chapter - 158
Casuarinaceae

Classification (Benthany and Hooker)

- Phanerogams
- Dicotyledons
- Monochlamydeae (or)
- Incompletae
- Unisexuals
- Casuarinaceae

General Characters:

→ Plants are evergreen profusely branched shrubs or trees. They possess long, slender, cylindrical and grooved branches. Uniqueness is they possess jointed or switch-like branches which bear alternating whorls of scaly leaves. These scale like leaves unite at the base forming a sheath.

→ Leaves - the number of leaves in a whorl varies from species to species and range from 4 to 12. The internodes are furrowed.

→ Flowers - unisexual and Perianth is absent

→ Male flowers - arise at the end of the branches and are found to be arranged in erect catkin like spikes. The flowers are arranged in whorls. Each whorl develops in the axil of a bract. This bract and other bracts situated at the node form a sheath which protects the young flowers. A pair of bracteoles and two small perianth leaves are also found. There is a single central stamen, showing the tendency to split. The filament is short and bears a two-celled anther which dehisce longitudinally. The filament elongates lengthwise before dehiscence and the perianth which first covers the anther is pushed off. The open anther is then pushed above the sheath.

→ Female flowers — are borne in dense heads at the end of short lateral branches. The flower is without perianth and consists of two carpels which are syncarpous. Ovary is bilocular but by the suppression of the posterior chamber it becomes unilocular. 2 ovules in parietal placentation. There are two long stigmas.

→ Fruit — one seeded winged nut protected by hardened bracteoles. Several fruits form a cone like structure.

→ Seeds — non endospermic; they are winged and being enclosed within woody bracteoles. Each seed bears a straight embryo with an upwardly directed radicle. Two large flat cotyledons.

→ Pollination is entamophilous.

Genera included under Casuarinaceae
- → Allocasuarina
- → Casuarina
- → Ceuthostoma
- → Gymnostoma

Classification — Moraceae (Bentham and Hooker)

           Phanerogams
           Dicotyledons
           Monochlamydeae
           Unisexuales
           Moraceae

General characters:

→ Mostly trees or shrubs

→ Herbs very rare (*Dorstenia*). Many species are epiphytic and sometimes form a tight network of roots round the stem of the host plant. The family is distinguished by the presence of latex, which is found in long sacs, especially in the secondary cortex or the phloem.

→ Leaves — Simple, alternate, rarely opposite, entire, serrate or lobed margins and stipulate.

→ Inflorescence — cymose or racemose. In *Ficus* sp, the cymes coalesce and form a fleshy hollow axis bearing flowers on the interior surface. Here the receptacles form a hollow cavity, with an apical opening possessing scales and inflorescence is known as hypanthodium. In each hypanthodium, the female flowers are found to be developed at the base of the cavity and the male flower near its apical region. In *Morus*, the inflorescence is catkin where the male and female flowers develop on different shoots.

→ Flowers :- unisexual, regular (actinomorphic), incomplete, small and hypogynous.

→ Perianth - Consists of four leaves which are occasionally united. They are arranged in 2 tepals. Perianth is persistent

→ Male flowers :- Number of stamens is equal to that of Perianth leaves and are situated opposite to Perianth leaves. Filaments incurved or straight. Anthers 2 celled, versatile and dehisce longitudinally. Some cases, stamen number is reduced to 1 or 2

→ Female flowers: - there are 2 medianly situated carpels, the hinder one of which shows various degrees of abortion. In many cases it is represented only by a style similar to that of the anterior carpel. In *Morus*, the hinder carpel is represented by empty ovary chamber or in some species it is developed. Sometimes the second style is represented by a small protuberance or altogether absent, and there is no remnant or indication of the hinder carpel (*Chlorophora*). The ovary is unilocular with a pendulous more or less curved ovule

→ Fruit - drupe, Perianth become fleshy & surrounds it. Some cases it is an achene. Some cases Sorosis, Synconus

→ Seed - Albuminous or exalbuminous with a bent or curved embryo.

Genera included under Moraceae

- Antiaris
- Antiaropsis
- Artocarpus
- Bagassa
- Batocarpus
- Bosqueiopsis
- Brosimum
- Broussonetia
- Castilla
- Clarisia
- Clarisia
- Craterogyne
- Cudrania
- Dorstenia
- Fatoua
- Ficus
- Helianthostylis
- Helicostylis
- Hullettia
- Maclura
- Maquira
- Mesogyne
- Metatrophis
- Milicia
- Morus
- Naucleopsis
- Olmedia
- Olmediopsis
- Parartocarpus
- Perebea
- Poulsenia
- Prainea
- Pseudolmedia
- Scyphosyce
- Sorocea
- Sparattosyce
- Streblus
- Treculia
- Trilepisium
- Trophis
- Trymatococcus
- Utsetela

Sub class: Monochlamydeae

Series: ordines anomali

Families:

Salicaceae

Lacsitemaceae

Empetraceae

Ceratophyllaceae

Chapter - 160
Salicaceae

Classification (Bentham and Hooker)

Phanerogams
Dicotyledons
Monochlamydeae
Ordines anomali
Salicaceae

General characters

→ Trees and shrubs
→ Leaves - alternate to opposite, spiral to distichous, flat, petiolate; non sheathing, simple, epulvinate lamina usually entire, rarely dissected; more or less exstipulate; if present, stipules are intrapetiolar. Margin - entire or dentate or crenate or serrate
→ Plants dioecious. Pollination - anemophilous or entomophilous
→ Inflorescence - catkins.
→ Flowers - individually bracteate, ebracteolate
→ Perianth absent or vestigial.
→ Androecium consists of 1 or 2 stamens or rarely 3-60 stamens; free of one another or coherent (monadelphous); Anthers - basifixed; dehiscing via longitudinal slits.

→ Gynoecium 2-4 carpelled; syncarpous, superior. Ovary unilocular. Placentation basal or parietal. 4-50 ovules per cavity. Anatropous
→ Fruit - non-fleshy, dehiscent, capsule
→ Seeds scantily endospermic or non endospermic

Genera included under Salicaceae:
→ Salix   → Populus   → Chosenia

Chapter - 161
Lacistemaceae

Classification (Bentham and Hooker)

Phanerogams
Dicotyledons
Monochlamydeae
Ordines anomali
Lacistemaceae

General characters

→ Shrubs and small trees
→ Leaves - alternate
→ Flowers - hermaphrodite
→ Inflorescence - catkins or racemes or panicles
→ Some species have sepals.
→ There is one stamen, one style and 2 stigmas
→ Fruit - capsule
→ One only seed

Genera included under Lacistemaceae

→ Lacistema
→ Lozania.

Genera included under Balanophoreae

- Balanophora
- Chlamydophytum
- Corynaea
- Dactylanthus
- Ditepalanthus
- Exorhopala
- Hachettea
- Helosis
- Langsdorffia
- Lophophytum
- Mystropetalon
- Ombrophytum
- Rhopalocnemis
- Sarcophyte
- Scybalium
- Thonningia

Chapter – 162
Empetraceae

Classification (Bentham and Hooker)
- Phanerogams
- Dicotyledons
- Monochlamydeae
- Ordines anomali
- Empetraceae

General characters

→ Small shrubs
→ Leaves - small, whorled or alternate, spiral; leathery, shortly petiolate, non-sheathing, simple, pulvinate. Lamina; entire; acicular or linear; exstipulate.
→ Plants dioecious or monoecious.
→ Flowers solitary or inflorescence - racemose
→ Flowers - bracteolate, regular, trimerous, tri to tetracyclic
→ Perianth with distinct calyx and corolla; or petaline or sepaline;
→ Calyx consists of 2-3 sepals, 1 whorled, polysepalous, regular, imbricate
→ Corolla - 2-3 petals, 1 whorled, polypetalous,

imbricate, regular. Petals shortly clawed to sessile.

→ Androecium 2-4 stamens, free of one another, 1 whorled; isomerous with perianth, oppositi sepalous, Anthers – extrorse or latrose; dehiscing via longitudinal slits.

→ Gynoecium 2-9 carpelled, syncarpous, superior; ovary 2-9 locular; styles-1, apical; stylar canal present. Stigma 1, lobed or 2-9. Placentation basal to axile. Ovules – 1 per locule; funicled;

→ Fruit – fleshy to non fleshy; indehiscent; drupe

→ Seeds endospermic

Genera included under Empetraceae

→ cenatiola → corema → Empetrum

Chapter - 163
Ceratophylleae

Classification (Bentham and Hooker)

Phanerogams
Dicotyledons
Monochlamydeae
Ordines anomali
Ceratophyllaceae

General characters

→ Submerged, brittle, rootless aquatic herbs.
→ Plants rootless
→ Hydrophytic, free floating
→ Leaves - submerged, whorled, 3-10 per whorl; becoming brittle with age; petiolate; simple or compound; lamina dissected; once or twice finely dichotomously dissected; leaves exstipulate.
→ Plants monoecious; pollination by water.
→ Flowers - solitary or inflorescence - axillary, ebracteate, ebracteolate, minute, regular, cyclic.
→ Perianth - sepaline (bractlike); 9-10, basally joined, 1 whorled
→ Androecium consists of 5-27 stamens; free of the

Perianth; subsessile anthers; anthers adnate; non-versatile; dehiscing irregularly or dehiscing via longitudinal slits.; extrorse.
→ Gynoecium —1 carpelled; monomerous, superior; stylate, apically stigmatic; 1 ovuled; Placentation apical; ovules pendulous.
→ Fruit - non-fleshy, indehiscent, achene
→ Seeds non-endospermic

Genera included under Ceratophyllaceae
→ Ceratophyllum

References

1. Taxonomy of Angiosperms – AVVS Sambamurthy
2. Modern Plant Taxonomy – N.S. Subramanyam
3. Taxonomy of Angiosperms – B.P. Pandey
4. Taxonomy of Angiosperms – Singh and Jain
5. Introduction to Principles of Plant Taxonomy – Sivaraja
6. Guide to Flowering Plant Families
7. Taxonomy of Angiosperms – Kumaresan
8. Wikipedia
9. www.wildflowers-and-weeds.com
10. www.botany.hawaii.edu
11. www.delta-intkey.com
12. www.theplantlist.org
13. www.wildflowers-and-weeds.com
14. www.britannica.com
15. www.biologydiscussion.com
16. en.m.wikisource.org
17. flora www.eeb.uconn.edu
18. www.babylon-software.com
19. https://eprints.utas.edu.au

www.ingramcontent.com/pod-product-compliance
Lightning Source LLC
Chambersburg PA
CBHW051340220526
45469CB00001B/49